MAPPING THE NATION

BUILDING A MORE RESILIENT FUTURE

Cover illustration by Daniel Gill, Esri

Esri Press, 380 New York Street, Redlands, California 92373-8100

Copyright © 2015 Esri
All rights reserved. First edition 2015.
Printed in the United States of America

19 18 17 16 15   1 2 3 4 5 6 7 8 9 10

CONTENTS

CONTENTS

FOREWORD

Resilient communities prepare ahead of time, operate effectively during a crisis, and recover quickly. Geographic information system (GIS) technology drives innovative applications—such as cloud-based services promoting collaboration among agencies—that help build more resilient communities. GIS helps community leaders anticipate future trends and enact policies that support rapid response during emergencies and disasters. GIS also supports long-range planning essential for resilience in such areas as health and social services, the economy, and food supply.

Mapping the Nation: Building a More Resilient Future, our annual federal map book, is filled with maps that convey the power and promise of GIS technology. Federal agencies apply GIS to a wide variety of solutions and workflows, such as targeting financial aid, combating the effects of climate change, protecting forests, and conserving water resources. The maps in this volume tell compelling stories of how GIS adapts to countless disciplines, engages citizens, and presents limitless opportunities to transform our nation and our planet.

I hope you enjoy the great work of the federal government featured in *Mapping the Nation* and continue to explore the exciting and innovative ways that geographic knowledge can be used and shared now and in the years ahead.

Warm regards,

Jack Dangermond
President, Esri

Learn more about the maps

Scan the QR codes throughout the book to access related websites and learn more about the maps and organizations behind them. (QR code readers are available at any app store.) Some content may require Adobe Flash Player.

INTRODUCTION

Focusing on transparency and resilience, the Obama administration has called for innovative ideas, tools, and policy improvements to make government smarter, more efficient, more collaborative, and more responsive to better serve the American people. At the forefront of the American Dream lies innovation, the driving force behind our nation's competitiveness. To change the path of government, innovation must go beyond great ideas and become enmeshed with the infrastructure in place. This can be done, in part, through GIS. Innovative maps and applications inspire creative ideas that lead to effective action that transforms not only government, but the world around us.

The Evolution of GIS

In the early-to-mid-1960s, people started working with something called computational geography. Roger Tomlinson in Canada coined the term "geographic information system" while Professor Howard Fisher at Harvard University and others were making computer maps. There was a shift from qualitative descriptions of places to the notion that we could put geographic data into a computer and do quantitative analytics, engaging theories of database management that were emerging in such fields as accounting.

Once you digitize data you can analyze patterns and relationships in geographic space between such things as health and pollution, plants and climate, and soils and landscapes. The applications moved quickly beyond the laboratory; businesses started getting engaged, and then cities realized they could have a centralized GIS database with updates on road changes, sewer lines, building permits, and land-use plans. States and national agencies started to embrace these same ideas for making better decisions, communicating more effectively, and encouraging departments to collaborate. The key word is integration. GIS organizes geography into layers covering sociology, geology, climatology, hydrology, and other disciplines. Engineers maintain their layer, planners theirs. GIS is the only technology that integrates many different subjects using geography as its common framework.

In the beginning, the technology was primarily used by GIS professionals and was unfamiliar territory to anyone without an extensive GIS background. Over the past few decades, GIS has evolved into a stronger, web-based platform for the GIS professional that also enables nontechnical users to understand how GIS can improve their organizations. This platform allows a range of users—from citizen to CIO—to leverage open data and better understand the location and relationships of demographics, economics, resources, and finances of the environment around them. Sharing data and creating simple web mapping applications with strong messages and narratives has never been easier.

The recent concept of location analytics is combining traditional tools such as business software suites and business intelligence solutions with organizational data, resulting in better-informed decisions in business and government. This part of the platform also expands the ways to use data, deliver maps, and perform analysis. As GIS technology becomes more accessible, elected federal officials are finding new ways to deliver government services using open data and a cloud-based platform. A nationwide push is encouraging the creation of startups and innovative ideas and tools using GIS to improve government efficiency, transparency, and community resilience.

Resilience is about sustainability, flexibility, and recovery through the study of such critical issues as climate change, infrastructure, public safety, social services, and the economy. Resilience engages citizens to become more informed about their communities.

Resilient Communities

Building more resilient communities requires the right tools. Governments across the globe have been using GIS to empower their organizations to be better equipped during challenging events, whether created by people or naturally occurring. Through this technology, community leaders can anticipate future trends and prepare for them.

Climate Resilience

Geospatial analysis is central to the White House Climate Action Plan by governments, nonprofits, and businesses to respond to climate-driven events, such as hurricanes, wildfires, and droughts. These organizations also use GIS to help develop sustainable urban transportation plans, generate clean energy, and increase energy efficiency.

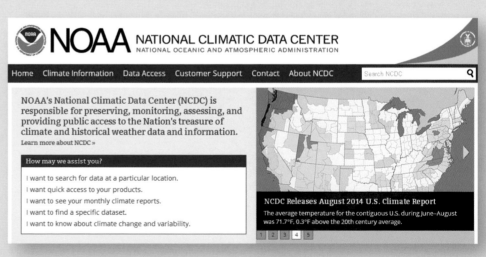

Document Link by climate_resilience.

 http://www.ncdc.noaa.gov/

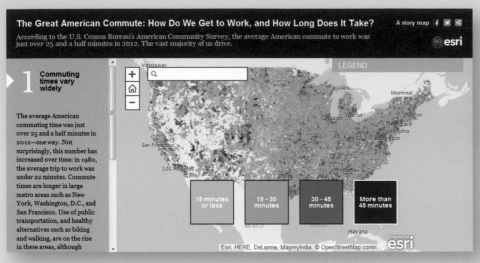

Web Mapping Application by Story Maps Team.

**http://storymaps.esri.
com/stories/2014/
commuting/**

Transportation and Infrastructure Resilience

To lessen commute times and reduce carbon emissions, transportation officials use GIS to analyze existing transportation infrastructure, areas of high demand, air quality, and opportunities for improvement. On a map, it becomes clear how a city can support a growing population with a multitude of options that effectively move people, promote economic growth and development, and improve air quality. GIS applications can connect citizens who want to carpool and can also identify the best routes for walking, biking, or taking public transit to work instead of driving.

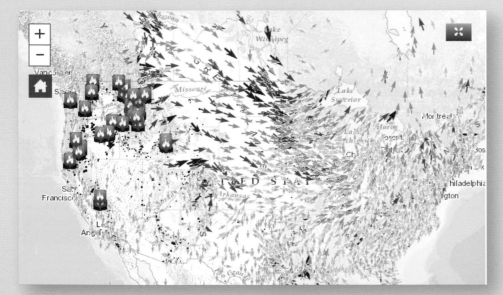

Web Map by esri_webapi.

http://esriurl.com/8487

Public Safety and Disaster Resilience

GIS helps government inform citizens about neighborhood risks and their impact. Apps enable people to create virtual neighborhood watches. GIS is vital to knowing where to send police officers or firefighters when a crime or disaster strikes and is changing the way public safety personnel respond to disasters. GIS can show leaders when and where to take preventative measures before floods, fires, and storms happen. After an event occurs, GIS helps response organizations see where the need for assistance is greatest, and where to quickly deploy teams to deliver aid. Citizen-generated apps help people see where they can donate time and resources.

Economic Resilience

The economy is constantly fluctuating and becoming increasingly tied to global events. During the recent economic recovery, resilient communities used GIS to better understand the changing landscape and the areas that were hit the hardest by the downturn. Moving forward, economic development departments use innovative approaches to help local business owners find the best ways to market their products and find the right places to set up shop.

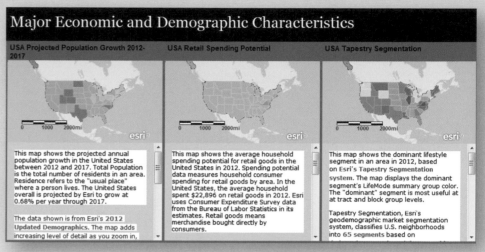

Web Mapping Application by Harry_Moore.

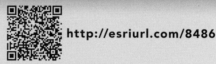

http://esriurl.com/8486

Food Resilience

The effects of a changing climate will have a significant impact on the world's food supply, so agricultural producers need tools like GIS to better understand and adapt to these risks. To grow crops and raise animals, farmers depend on specific climate conditions. As these conditions change, GIS helps farmers meet these challenges.

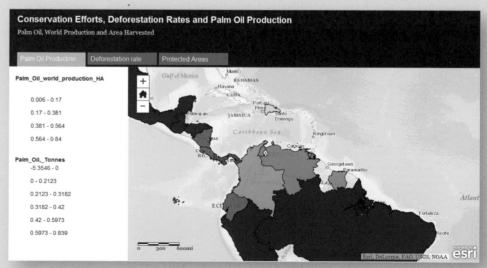

Application by p.desalvo_geoss.

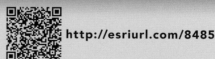

http://esriurl.com/8485

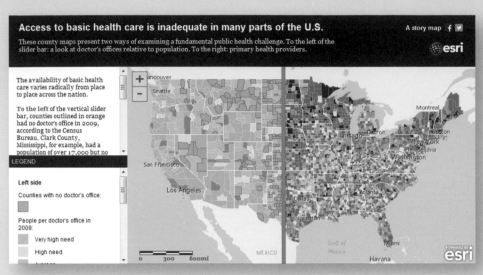

Web Mapping Application by Story Maps Team.

http://storymaps.esri.
com/stories/doctors/

Health and Social Service Resilience

Resilient communities use GIS to determine where to build clinics and hospitals that serve changing and aging populations as well as how to prepare for potential outbreaks. During extreme weather events, governments use GIS to determine community needs, such as where to set up cooling stations in a heat wave. Social services organizations rely on GIS to analyze transportation networks, service sites, demographics, and other layers of essential information for assessing, delivering, and integrating their programs.

Web Map Application by Bernie S.

http://esriurl.com/8484

Global Resilience

As communities around the world develop and concentrate wealth and population into focused urban areas, their vulnerability to disasters increases. The United Nations Office for Disaster Risk Reduction has launched a global campaign aimed at making communities more resilient with a goal of reducing political and economic instability from disasters. Geospatial technology serves as a key platform to connect cities, governments, and private organizations working to assess risk exposure and increase overall resiliency.

Enhancing Civic Policy and Engagement

Elected officials face the constant challenge of explaining complex and intricate policies to constituents. In turn, constituents want answers to important questions such as these: How is a policy going to affect our community? How is legislation helping to bring economic development to our district? Has funding been used wisely? For the myriad policy decisions that legislators face on a daily basis, they rely on accurate and authoritative data to keep citizens informed.

In our rapidly changing digital world, citizens are now demanding more from elected officials. They are conducting more government business online by signing online petitions, e-mailing legislators, and leveraging social media as an advocacy tool. This digital ecosystem has created a new frontier of civic engagement. Congress has begun to adopt emerging technologies to meet their citizens in this digital space. Although many interactions take place online, one constant remains: nearly every government process is tied to location. This means that GIS is uniquely positioned to transform the way Congress arrives at policy decisions and engages with citizens. For legislators, GIS offers an opportunity to provide clarity from complexity and help government craft smarter policies.

GIS as an Integrative Platform

GIS adoption by Congress is still emerging, but the foundation for exponential growth has been set. For legislators, GIS comes at a time when society is creating more data than ever before. For many across government, GIS has become an integrative technology, providing the ability to leverage different data sources and different technologies (cloud, big data, mobile), and increased awareness to events in real time.

In the past few years, the technology has evolved and extended across federal agencies. Today, GIS tools easily create basic maps and GIS users no longer need to be trained professionals. Although GIS professionals continue to play an important role, cloud computing and the development of easy-to-use web services have extended the base of users. Policy makers can connect to a map and quickly see the impacts of their policy. Ultimately, GIS provides the opportunity for better policy and improved outcomes for citizens.

Supporting a Digital Future

Today's world requires powerful tools such as GIS to help government anticipate future trends, plan wisely, and solve complex problems. Once legislators see how GIS analyzes data, saves time, and reduces costs, they can use the technology to justify policy decisions and provide clarity to constituents. The maps in this volume provide a glimpse at how government can promote innovation and resilience by supporting a digital future.

"What I need from (GIS professionals) is curated, authoritative, timely data in order to make legislative decisions. I need the best you have available so that we can do things that make sense."

CATHY CAHILL

Legislative Fellow, Senate Committee on Energy and Natural Resources

US DEPARTMENT OF AGRICULTURE

Early Access to National Agriculture Imagery Program Digital Files

Farm Service Agency (FSA) county staff needs National Agriculture Imagery (NAIP) files delivered as soon as possible. The standard delivery of NAIP digital imagery took two to four months, but the timeline needed to be improved to better support FSA programs. NAIP contractors developed a delivery service that made imagery available within five business days of acquisition, loaded into ArcGIS Online, and shared. FSA county staff now has access to the NAIP imagery between eight and fourteen weeks earlier than the standard delivery mechanism. This has resulted in better decisions on tighter timelines.

http://esriurl.com/8465

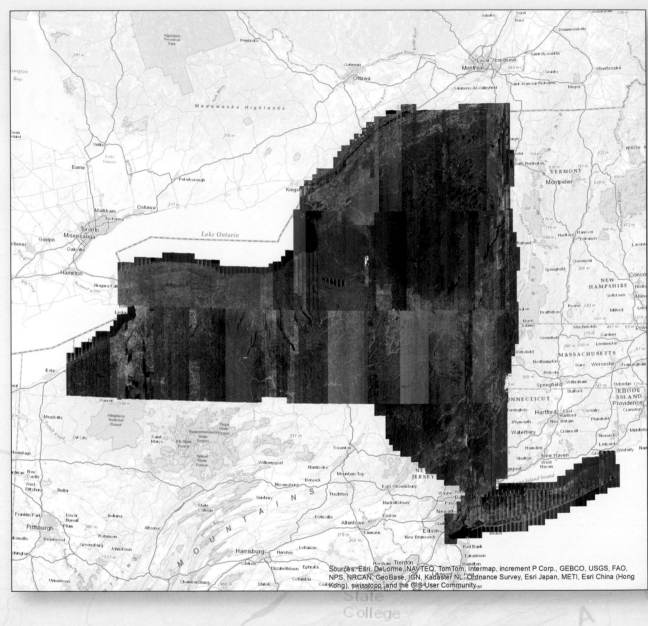

Sources: Esri, DeLorme, NAVTEQ, TomTom, Intermap, increment P Corp., GEBCO, USGS, FAO, NPS, NRCAN, GeoBase, IGN, Kadaster NL, Ordnance Survey, Esri Japan, METI, Esri China (Hong Kong), swisstopo, and the GIS User Community

Early Access to National Agriculture Imagery Program Digital Files, continued

Number of Days between Early Access and Standard Deliveries

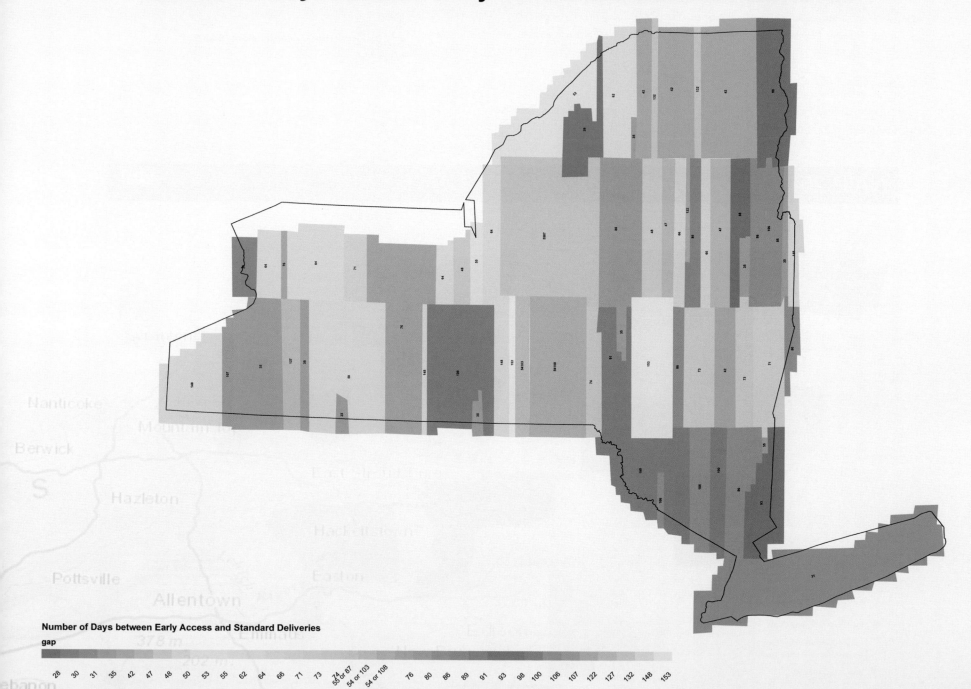

Number of Days between Early Access and Standard Deliveries

gap

28 30 31 35 42 47 48 50 53 55 62 64 66 71 73 74 55 or 87 54 or 103 54 or 108 76 80 86 89 91 93 98 100 106 107 122 127 132 148 153

National Agriculture Imagery Program 2013 Feedback

The National Agriculture Imagery Program (NAIP) gathered user feedback about its imagery through a text-based e-mail system. The system lacked an easily accessible spatial component and users could not view feedback from the user community. A pilot project using an ArcGIS Online public map and the public NAIP image services placed user feedback on a web map. The feedback map is displayed on an Internet browser, so users can access it on desktop computers, tablets, and smartphones. With the ArcGIS Online map, feedback is more accessible to the public, users are not required to have access to specialized software, information is gathered from a diverse knowledge base, the information is spatial, feedback is viewed as a whole, and NAIP management makes better-informed decisions.

 http://esriurl.com/8466

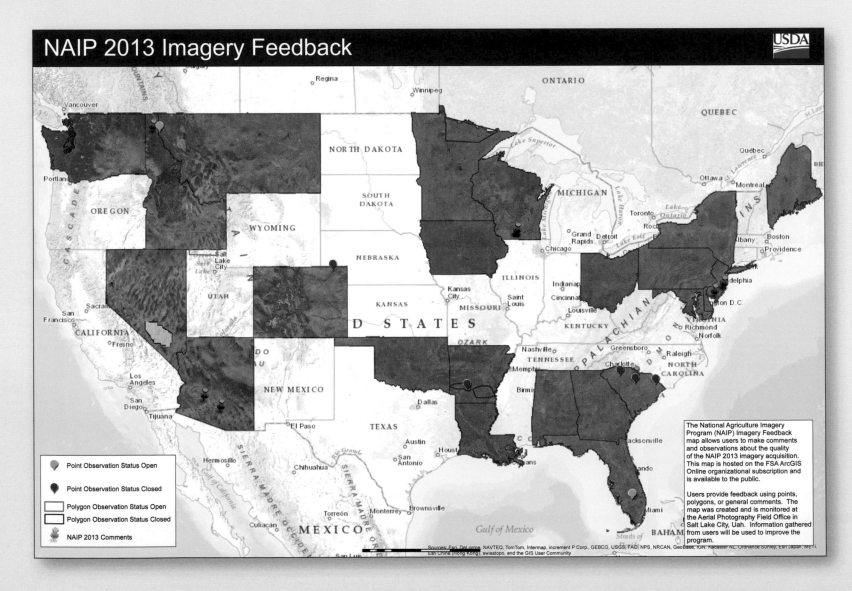

Bighorn National Forest Wildfires 1900–2013

Bighorn National Forest Wildfires 1900 - 2013

The US Forest Service wanted to provide information to high school students on the distribution of wildfires in size, time, and space across the Bighorn National Forest. A poster-sized map displays the extent of each fire and uses color coding to identify the time period. This helps high school students and others understand that wildfires vary in the size and number that occur per year and that fires are distributed across the entire forest landscape.

http://www.fs.usda.gov/
bighorn

Wyoming

Large Fires > 5 acres since Jan 1, 2000
Large Fires > 5 acres before Jan 1, 2000
Small Fires < 5 acres since Jan 1, 2000
Small Fires < 5 acres before Jan 1, 2000
Bighorn NF
Highways

Forest Health Maps

A wide range of insects and diseases kills trees across the nation's forests. Natural resource agencies, legislators, scientists, and the public want a hazard assessment of forest health concerns. The Forest Service coordinates a GIS modeling process to periodically update the *National Insect and Disease Risk Map*, a nationwide strategic assessment of the hazard of tree mortality from insect and disease from 2013 to 2027. GIS allows the US Forest Service to model and map potential forest insect and disease hazards. The maps shown here focus on the declining whitebark pine and red bay trees. This assessment provides a scientific basis to strategically allocate forest health protection resources across regions and individual pest distributions.

**http://www.fs.fed.us/
foresthealth/technology/**

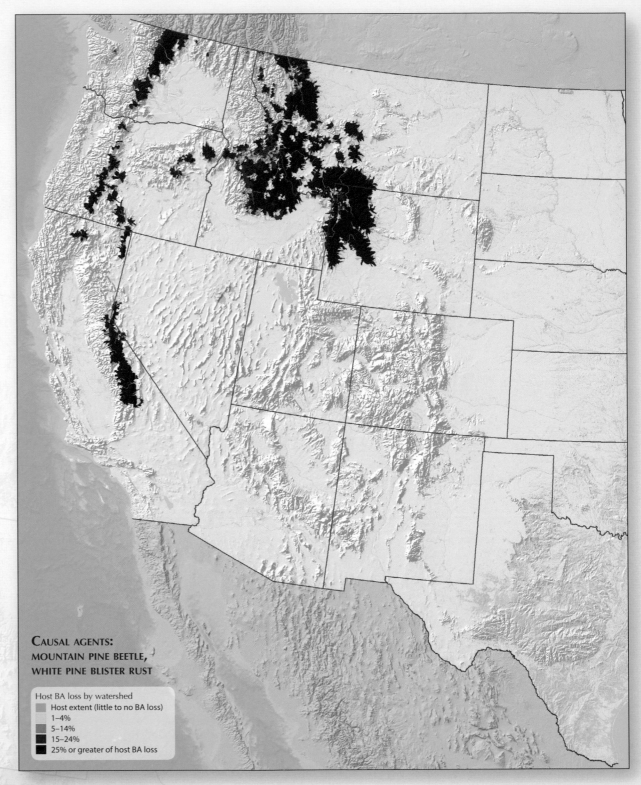

**CAUSAL AGENTS:
MOUNTAIN PINE BEETLE,
WHITE PINE BLISTER RUST**

Host BA loss by watershed
- Host extent (little to no BA loss)
- 1–4%
- 5–14%
- 15–24%
- 25% or greater of host BA loss

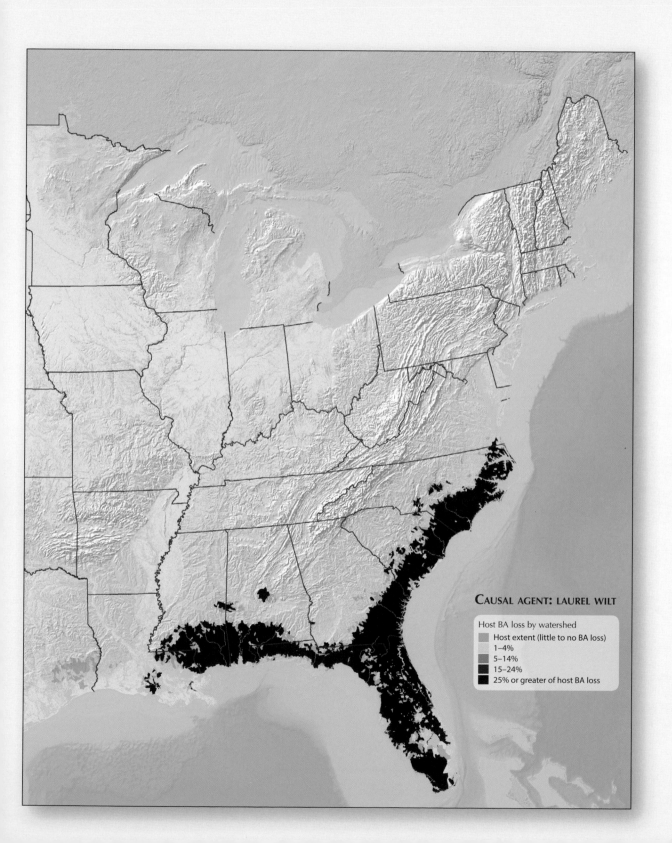

CAUSAL AGENT: LAUREL WILT

Host BA loss by watershed
Host extent (little to no BA loss)
1–4%
5–14%
15–24%
25% or greater of host BA loss

2012 Census of Agriculture Atlas

The agriculture census, with more than six million pieces of information, is conducted every five years to show how US agriculture is changing, what is staying the same, what is working, and what can be done differently. An atlas of 248 maps, including the examples shown here, was created for the 2012 agriculture census to make the data accessible and easily understood. This information is used by everyone who provides services to farmers and rural communities including federal, state, and local governments, and agribusinesses.

http://esriurl.com/8467

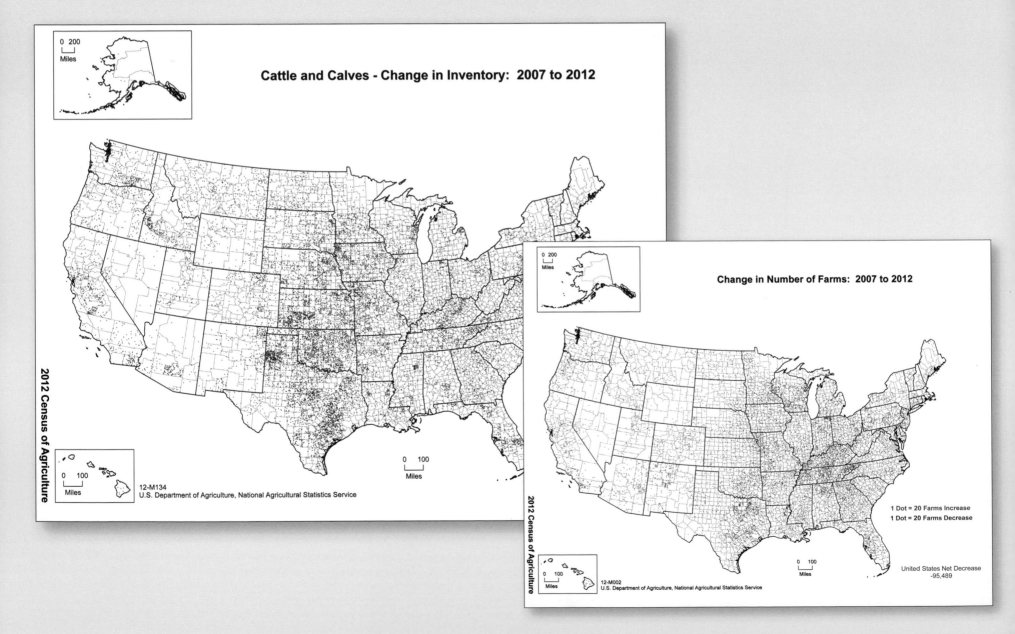

Cattle and Calves - Change in Inventory: 2007 to 2012

2012 Census of Agriculture

12-M134
U.S. Department of Agriculture, National Agricultural Statistics Service

Change in Number of Farms: 2007 to 2012

1 Dot = 20 Farms Increase
1 Dot = 20 Farms Decrease

United States Net Decrease
-95,489

12-M002
U.S. Department of Agriculture, National Agricultural Statistics Service

2012 Census of Agriculture Atlas, continued

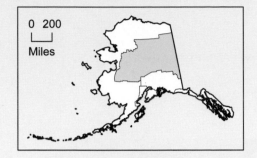

Expenses for Chemicals
as Percent of Total Farm Production Expenses: 2012

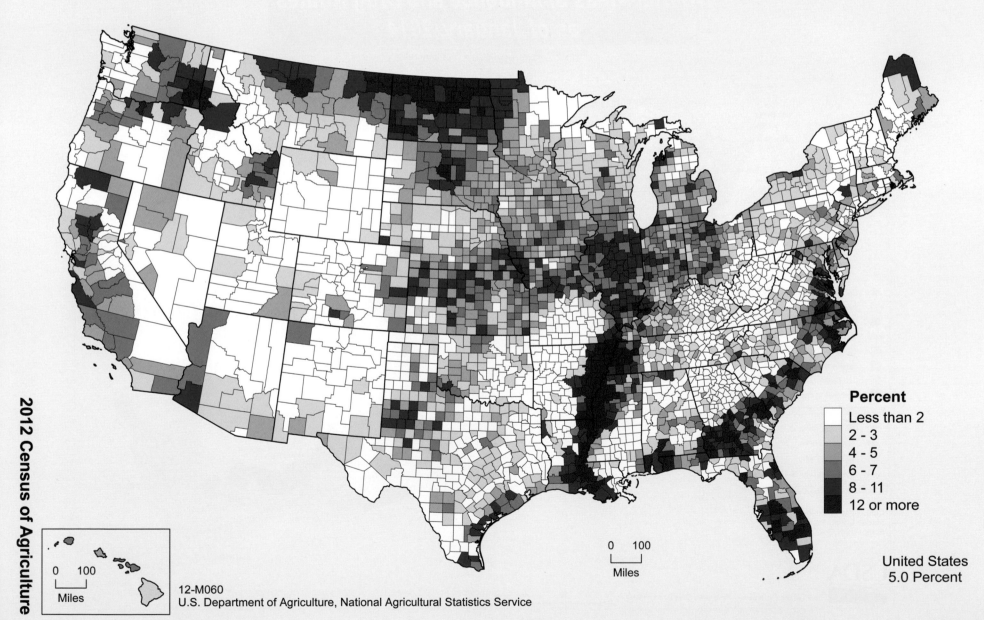

Percent
- Less than 2
- 2 - 3
- 4 - 5
- 6 - 7
- 8 - 11
- 12 or more

United States
5.0 Percent

2012 Census of Agriculture

12-M060
U.S. Department of Agriculture, National Agricultural Statistics Service

Ports on Mexico-US Border and Tracking the Drug Route

Many US Department of Agriculture (USDA) employees who work along the Mexican border or cross the border for food inspection are at risk from drug traffickers. The border project was developed to track drug routes and USDA office locations using maps for the border area and interior of Mexico. The maps have helped decision makers set up a broad plan to ensure USDA employees' safety.

http://www.dm.usda.gov/ohsec/

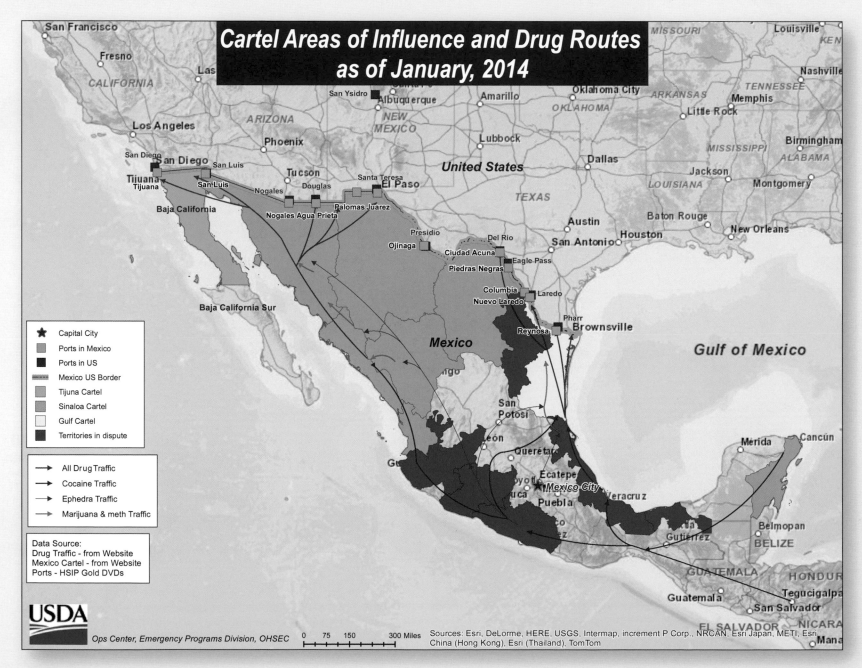

2013 Inauguration Support

http://www.dm.usda.gov/ohsec/

Many US Department of Agriculture (USDA) facilities and employees around the Capitol Hill area in Washington, DC, were affected by the January 2013 presidential inauguration. The USDA headquarters hosted over a thousand National Guard troops. The USDA Office of Homeland Security and Emergency Coordination provided inauguration support, including large maps that helped decision makers successfully plan and coordinate the big event.

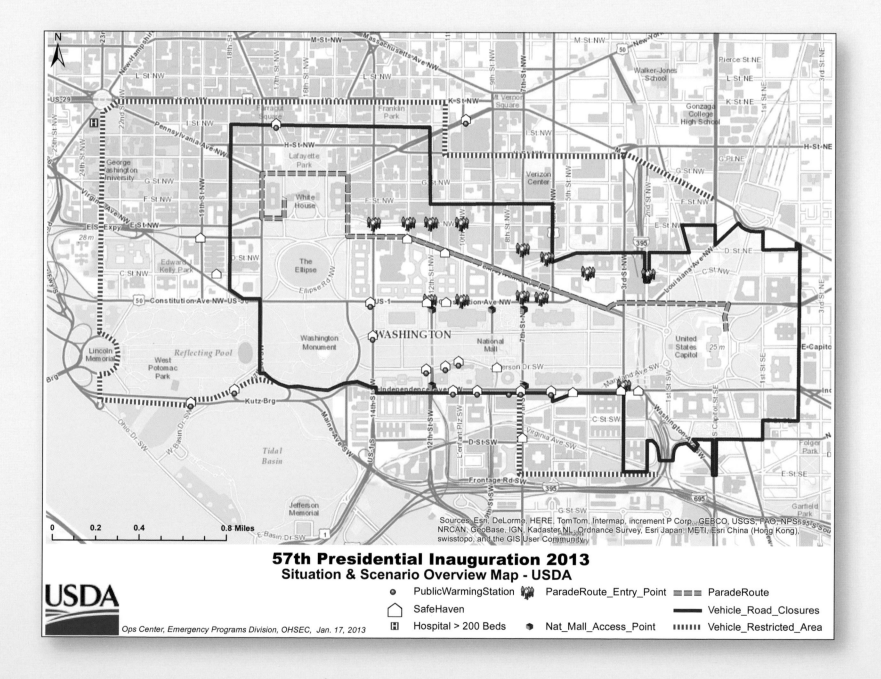

57th Presidential Inauguration 2013
Situation & Scenario Overview Map - USDA

Ops Center, Emergency Programs Division, OHSEC, Jan. 17, 2013

● PublicWarmingStation	👥 ParadeRoute_Entry_Point	▪▪▪ ParadeRoute	
⌂ SafeHaven		▬ Vehicle_Road_Closures	
Ⓗ Hospital > 200 Beds	● Nat_Mall_Access_Point	▪▪▪▪▪ Vehicle_Restricted_Area	

Sources: Esri, DeLorme, HERE, TomTom, Intermap, increment P Corp., GEBCO, USGS, FAO, NPS, NRCAN, GeoBase, IGN, Kadaster NL, Ordnance Survey, Esri Japan, METI, Esri China (Hong Kong), swisstopo, and the GIS User Community

US DEPARTMENT OF COMMERCE

Comparing Metro and Micro Area Population Change between 2002–2003 and 2012–2013

Detecting growth and population declines in metropolitan and micropolitan statistical areas can be an unwieldy task for some census data users. This story map simplifies "big data" with a tool that enables users to visualize population growth for two time periods with an interactive swipe bar. The side-by-side maps show percentage change in population for all metro and micro areas for both 2002–2003 (left map) and 2012–2013 (right map). Pop-up boxes containing the area's title and its numeric and percentage change values for each period are another interactive feature of this template. In an instant, these maps reveal the fastest- and slowest-growing groups of metro and micro areas across the nation.

http://esriurl.com/8468

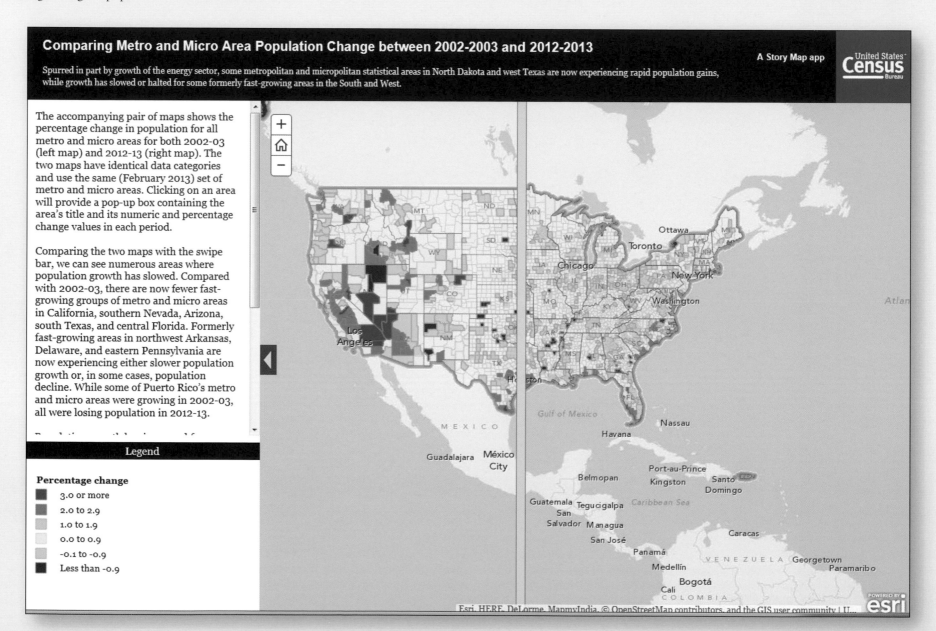

County Population Growth between 2012 and 2013 and the Primary Source of Population Change

http://esriurl.com/8469

Maps that tell a story, such as population growth, are not new to the US Census Bureau, but demographers now use interactive story maps to transform statistics into easily understood images. These maps use a slider to compare the growth in population for all counties in the United States (left map) and the component that contributed the most to growth within each county (right map). Pop-ups show population changes for each county.

County Population Growth Between 2012 and 2013 and the Primary Source of Population Change

A Story Map app

United States Census Bureau

Between 2012 and 2013, migration was the primary driver of change in the high growth counties, particularly in counties in gas- and oil-rich areas in or around the Great Plains.

Moving the slider will allow you to switch between seeing percent change and the primary source of growth in counties that are gaining population.

These maps show the percent growth in population for all counties in the U.S. **(left map)** and the component that contributed the most to growth within the county **(right map)**. Both maps are based on the most recent set of county population estimates released by the Census Bureau.

Clicking on the percent growth map will provide a pop-up box containing the county name, the county's 2013 population, percent population change from 2012 to 2013, as well as the change in the components of the estimates from 2012 to 2013. The components of estimates change are natural increase (both births and deaths) and migration (both net domestic and net international migration).

Clicking on the component of growth map will provide a pop-up box indicating which

Legend

Percent growth
- 3.0 or more
- 2.0 to 2.9
- 1.0 to 1.9
- Up to 0.9
- No growth

Primary source
- Natural increase
- Net migration
- Equal
- No growth or no primary source

Esri, HERE, DeLorme, MapmyIndia, © OpenStreetMap contributors, and the GIS user community | U...

POWERED BY esri

15

Metropolitan and Micropolitan Statistical Areas of the United States and Puerto Rico

The federal Office of Management and the Budget collects, tabulates, and publishes statistics for geographic areas to carry out government programs and allocate funds. Maps show that major urban populations of the United States are primarily located within Metropolitan Statistical Areas (greater than 50,000 population) and Micropolitan Statistical Areas (greater than 10,000 and less than 50,000 population). These two types of statistical areas are the component counties of Core Based Statistical Areas and Metropolitan Divisions (2.5 million population). Metropolitan and micropolitan statistical areas comprise the central county or counties or equivalent entities containing the core, plus adjacent outlying counties having a high degree of social and economic integration with the central county or counties as measured through commuting.

http://www.census.gov/geo/
maps-data/maps/cbsacsa.html

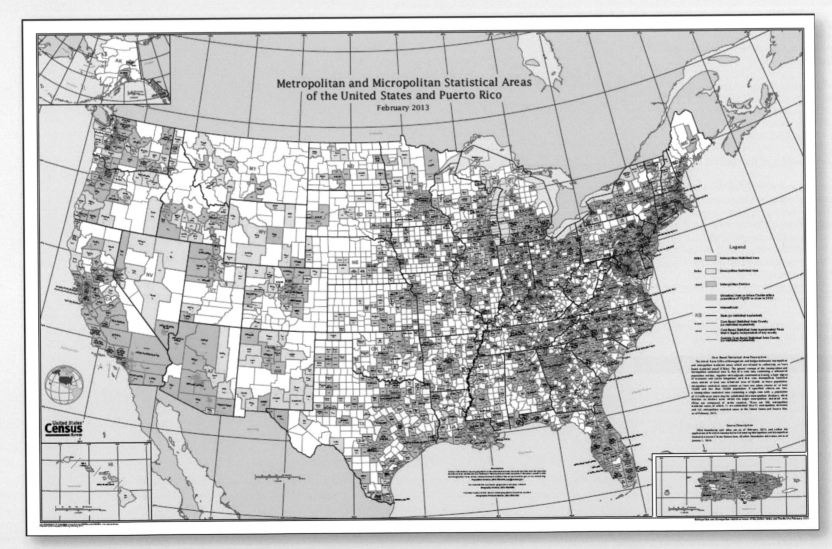

New England City and Town Areas

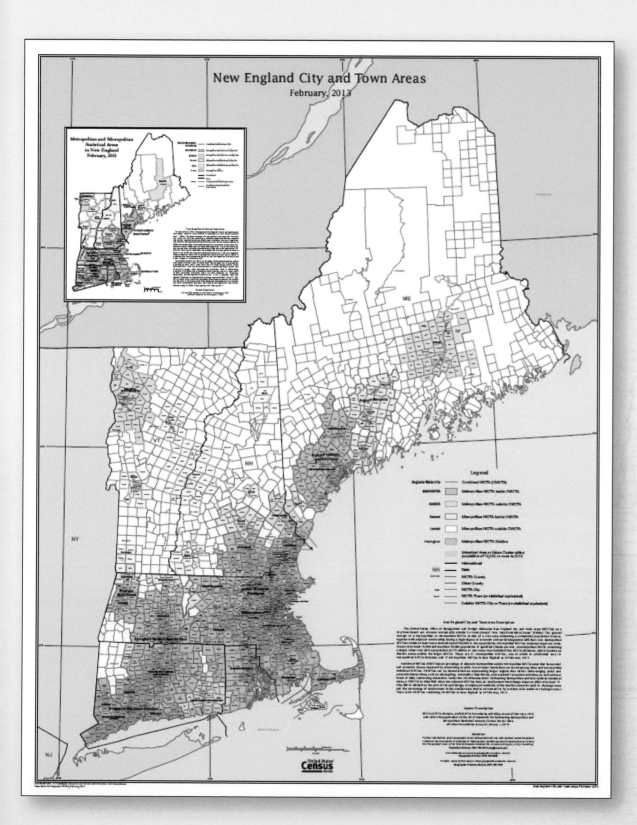

In New England, towns are a much more important level of government than counties, so the federal government established the New England City and Town Area (NECTA) as a geographic and statistical entity for the six-state New England region of the United States. NECTAs are defined using the same criteria as county-based Core Based Statistical Areas (CBSAs) and, similar to CBSAs, NECTAs are categorized as metropolitan or micropolitan.

http://www.census.gov/
geo/maps-data/maps/
nectas.html

PUMA Reference Map: San Bernardino County and Redlands and Yucaipa Cities

Public Use Microdata Areas (PUMAs) partition each state into areas containing about 100,000 residents and provide the most detailed geographic areas available in samples collected for the American Community Survey. The nationwide survey is designed to provide more frequently updated demographics for national and subnational geography than provided by the decennial census program. This PUMA map focuses on the San Bernardino County, California, cities of Redlands and Yucaipa.

 https://www.census.gov/geo/reference/puma.html

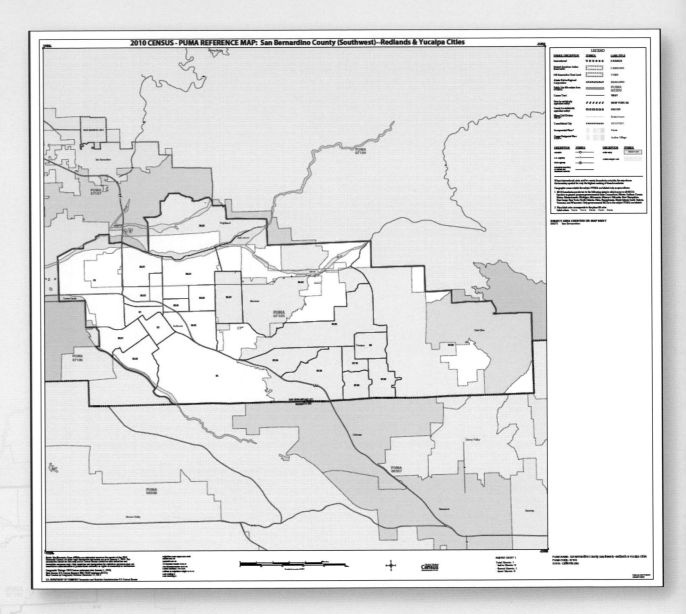

SAIPE Derived Poverty Surface: Poverty Ratio for Related Children Ages 5 to 17

SAIPE Derived Poverty Surface

Poverty ratio for related children ages 5 to 17

2012

Poverty Ratio (%)
>50
37.5
25
12.5
0

The government needs to track child poverty to administer federal programs and allocate federal funds to local jurisdictions. The Small Area Income and Poverty Estimates (SAIPE) program uses GIS to visualize child poverty in the United States as a continuous smoothed surface that cuts across state boundaries. SAIPE calculates the child poverty ratio for unified and elementary school districts for each state using model-based estimates of income and poverty statistics combined with administrative records, population estimates, census data, and estimates from the American Community Survey. State and local programs also use the income and poverty estimates for distributing funds and managing programs.

http://esriurl.com/8470

High-Resolution Coastal Change Analysis Program

Knowing how and where land cover is changing is essential for understanding impacts from past practices and setting the right course for the future. But land cover data is often inconsistent and can be difficult to use or interpret. The Coastal Change Analysis Program of the National Oceanic and Atmospheric Administration (NOAA) addressed that need by producing land cover and land change information for the coastal regions of the United States. These nationally standardized datasets are accessible in the Land Cover Atlas, a user-friendly online viewer that does not require desktop GIS software or advanced technical expertise. Maps focusing on Hawaii and Horry County, South Carolina, are shown here. Through the Land Cover Atlas, NOAA is providing greater access to this data, summarizing general change trends, highlighting specific items of interest, and providing users with printable reports.

 http://www.csc.noaa. gov/digitalcoast/data/ ccapregional/

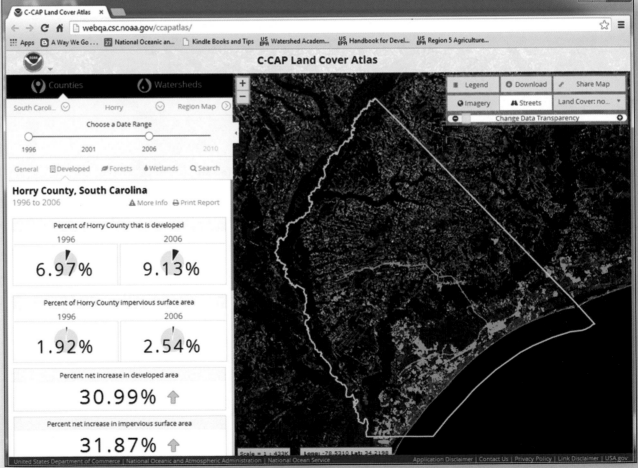

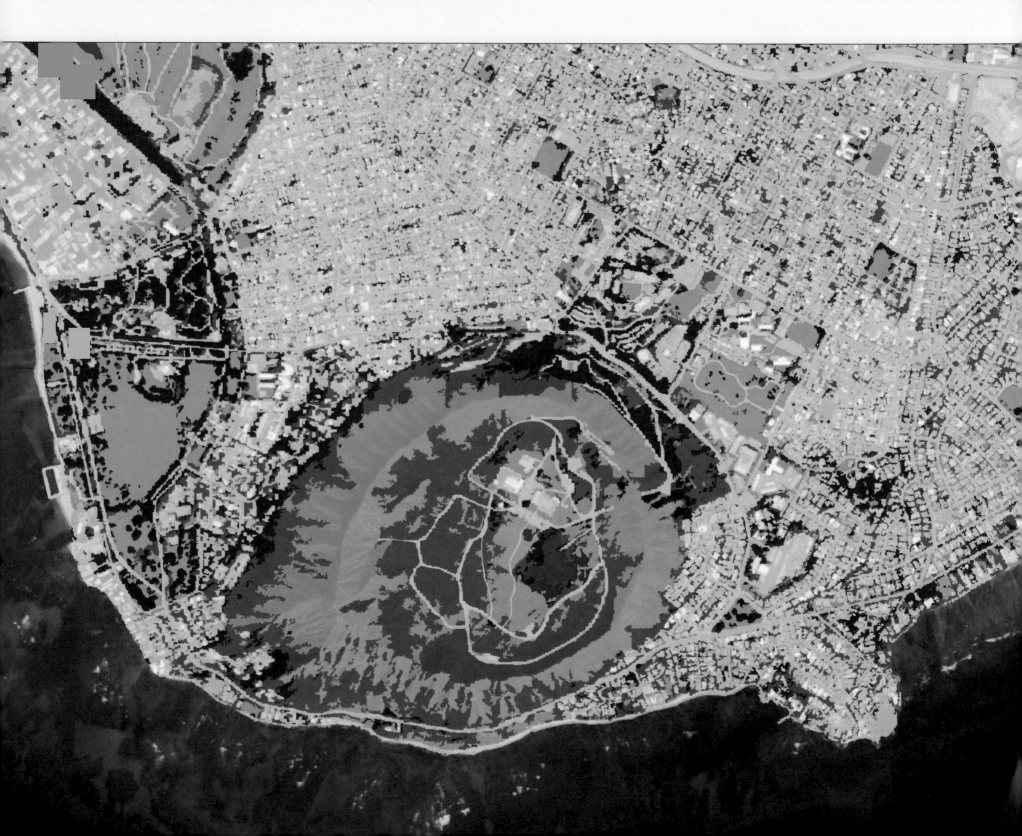

Washington State Spatial Prioritization Data Viewer

Inconsistent information on seafloor mapping surveys hindered marine planning strategies to minimize coastal zone conflict and reduce human-induced impacts to ecological resources. This data viewer provides users access to a map-based inventory, including type, extent, vintage, and quality of seafloor mapping surveys available along Washington state's outer coast. A unified collection of disparate seafloor data supports planning for future projects by minimizing duplication and maximizing reuse of existing data.

 http://esriurl.com/8471

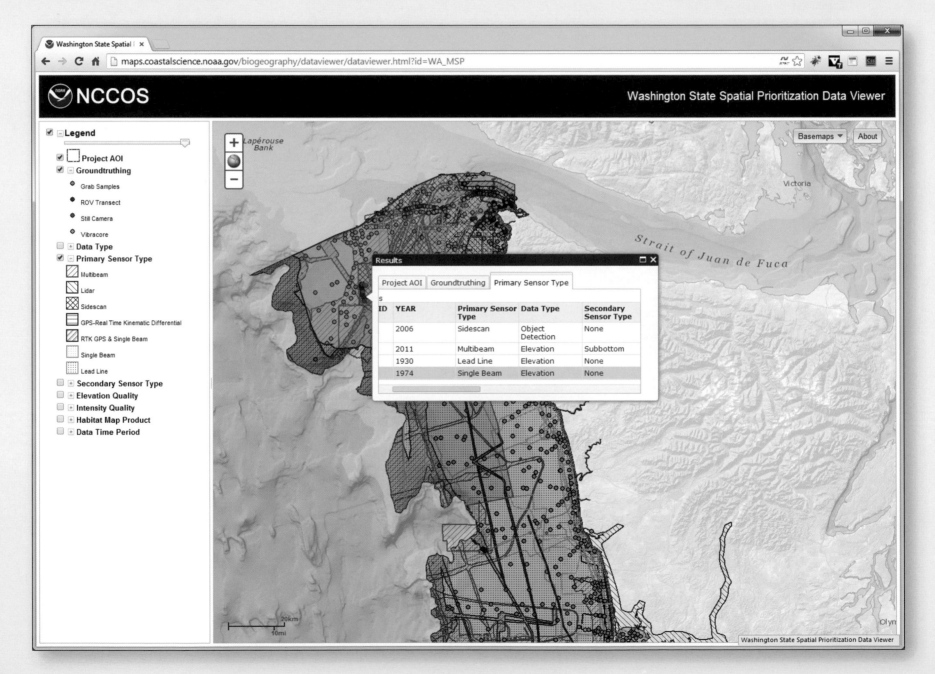

Mapping America's Coral Reefs

http://maps.coastalscience.noaa.gov/coralreef/#

The National Centers for Coastal Ocean Services wanted an accessible, understandable format for a government report on how marine resource managers use seafloor habitat maps. A story map was designed to explore how maps help communities study and protect coral reef ecosystems. *Mapping America's Coral Reefs* integrates beautiful imagery, maps, and first-hand accounts from experts in a single location and for a broad audience, from legislators to the general public.

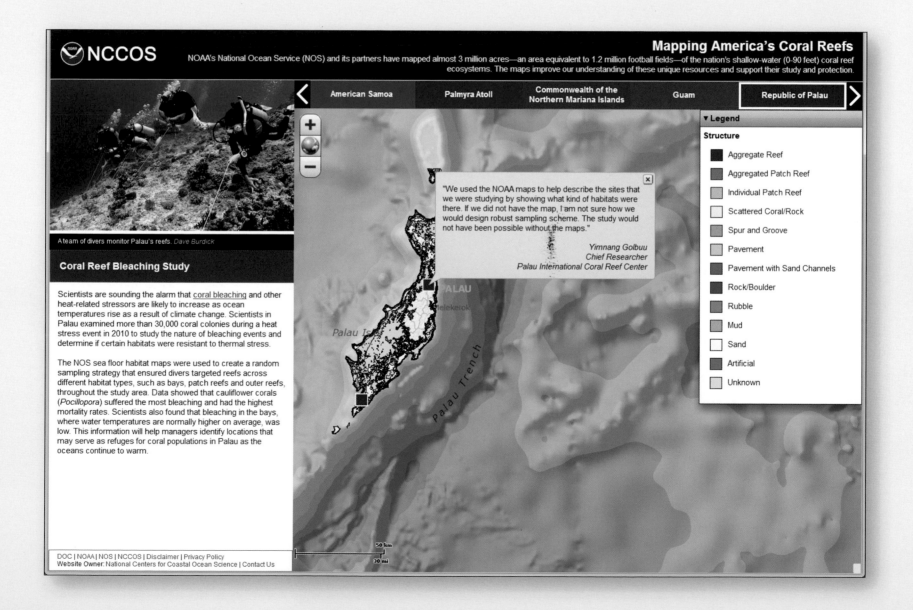

Estimated Changes New Datums Will Create

A geodetic datum is an abstract coordinate system with a reference surface (such as sea level) that serves to provide known locations to begin surveys and create maps. In 2022, the National Oceanic and Atmospheric Administration's National Geodetic Survey plans to introduce new national geometric and geopotential (vertical) datums to define the US National Spatial Reference System. They will provide more consistent three-dimensional positioning throughout the United States and its territories, including changes in position with time. The new datums will replace both the North American Datum of 1983 (NAD 83) and the North American Vertical Datum of 1988 (NAVD 88). These maps provide approximate predicted horizontal and vertical changes between the current datums and the new datums.

 http://www.ngs.noaa.gov/

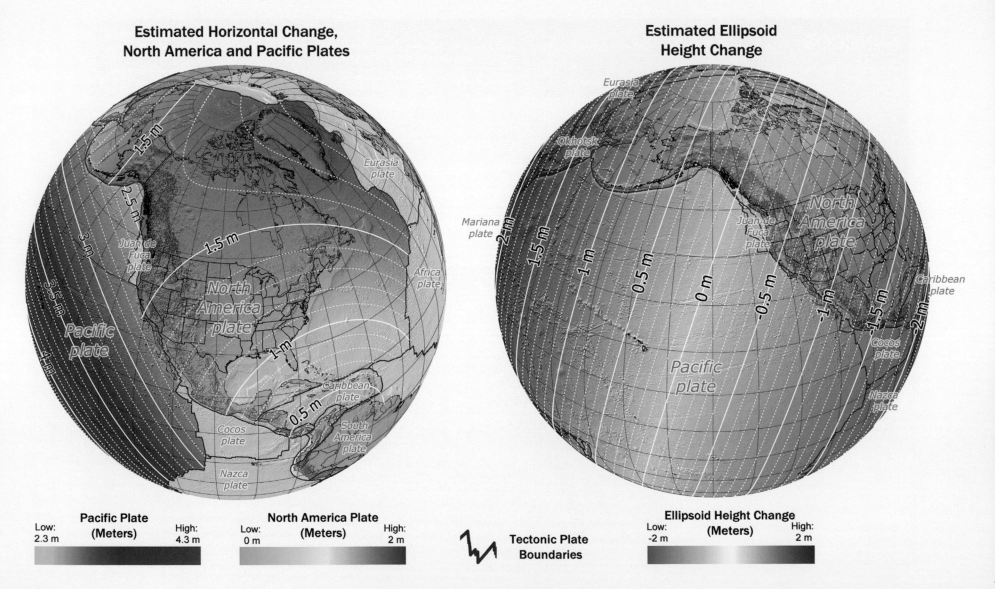

Estimated Horizontal Change, North America and Pacific Plates

Estimated Ellipsoid Height Change

Pacific Plate (Meters)		
Low: 2.3 m		High: 4.3 m

North America Plate (Meters)		
Low: 0 m		High: 2 m

Tectonic Plate Boundaries

Ellipsoid Height Change (Meters)		
Low: -2 m		High: 2 m

Natural Hazards Viewer

http://maps.ngdc.noaa.
gov/viewers/hazards/

Natural hazards such as earthquakes, tsunamis, and volcanoes affect both coastal and inland areas. Long-term data from these events can be used to establish the past record of natural hazard events, which is important in planning for, responding to, and mitigating future events. Shown here on the Natural Hazards Viewer are historical tsunami sources surrounding Japan. By carefully analyzing historical and real-time data, scientists can use hurricane satellite images to help protect property and reduce the injuries, trauma, and deaths that accompany a natural hazard event. Rigorous monitoring and studying of geological and meteorological conditions has enabled the United States to develop some of the best predictive capabilities in the world.

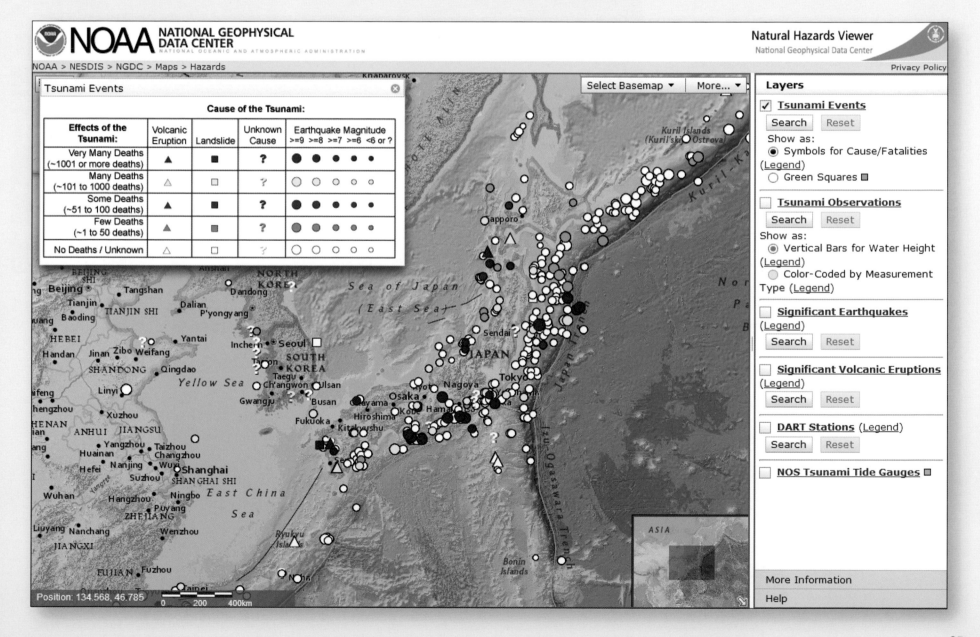

Hillside Visualizations of National Ocean Service Bathymetric Attributed Grids

The National Geophysical Data Center archives high-resolution seafloor elevation data from hydrographic surveys in US coastal waters. The data is available as bathymetric attributed grids, and many areas have been mapped at very high detail. Color-shaded relief images offer a way to assess data coverage and quality before downloading and also provide beautiful and engaging basemaps in combination with other map layers. The images shown here include coastal areas of Northern California, Washington, and Alaska. Such images enable scientists to model tsunami propagation and ocean circulation and assist in seafloor habitat research, weather forecasting, and environmental stewardship.

http://maps.ngdc.noaa.gov/
viewers/bathymetry/

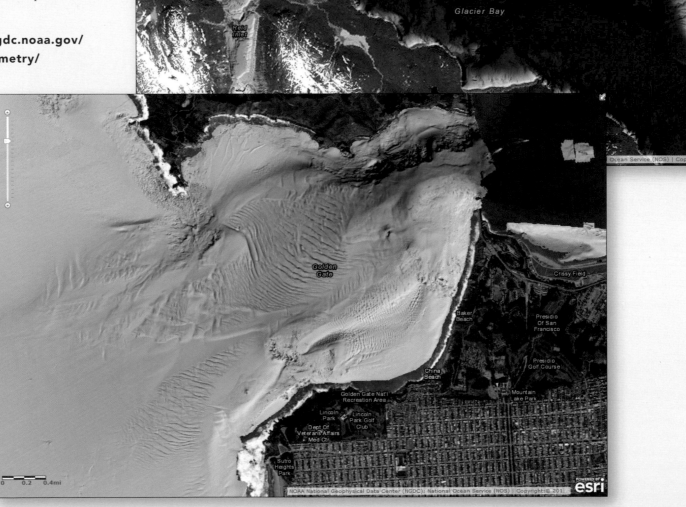

Hillside Visualizations of National Ocean Service Bathymetric Attributed Grids, continued

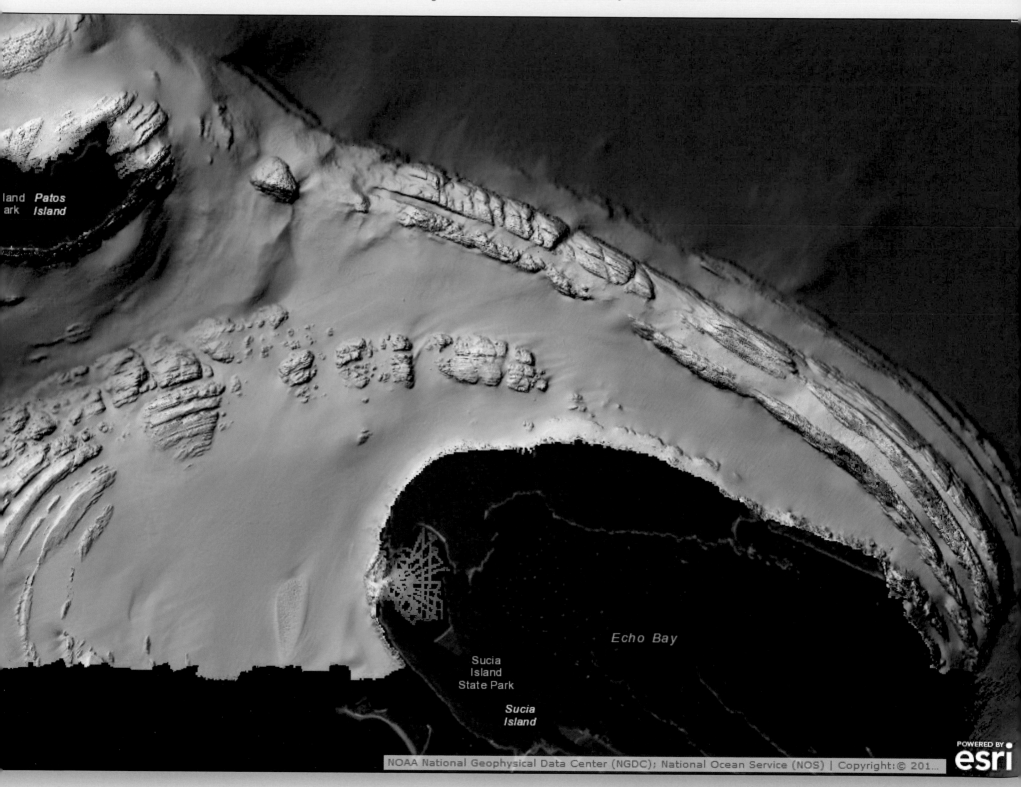

Hurricane Sandy Reconstruction

Addressing flood risk based on current conditions has immediate, short-term benefits to communities but does not adequately account for increasing flood risk resulting from sea-level rise. Using the best available science and data, federal agencies jointly developed a tool to help state and local officials understand possible future flood risks from sea-level rise. This tool provided guidance for post-Sandy rebuilding by supporting scenario planning to help officials adapt to uncertainties. The tool also showed risk associated with projected scenarios of sea-level rise.

 http://esriurl.com/8472

National Weather Service Maps (Highs, Lows, Current Temperatures)

The National Weather Service (NWS) did not have any nationwide interpolated form of observed high and low temperatures. Using GIS and observed weather data from across the country and around the world, the NWS created a series of current weather maps, including temperatures and dew points, that update every hour. These maps are an easy and understandable portrayal of current temperatures from across the United States. This is beneficial for communicating current and forecast weather information.

http://www.weather.gov/btv/

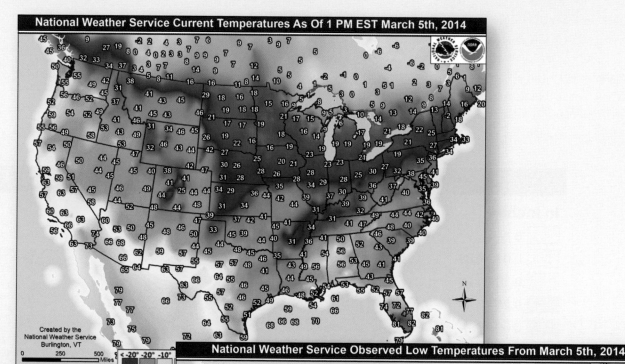

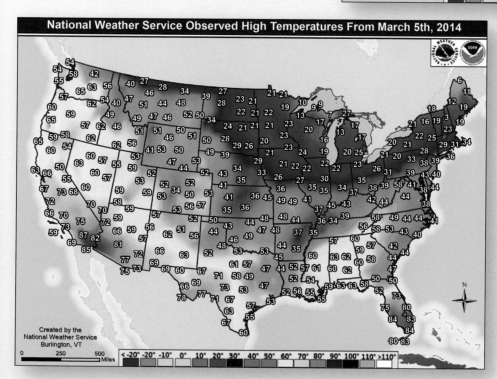

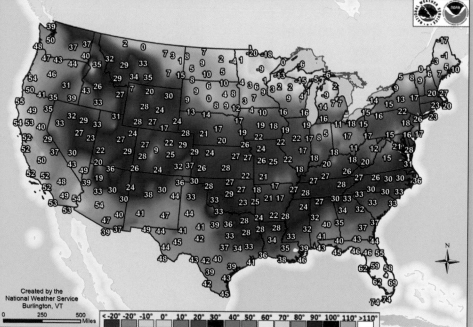

National Weather Service Total Precipitation Forecast

The forecast maps of the National Weather Service (NWS) were not user friendly, and forecasts were not always easy to communicate. There was a severe lack of regional and national mapping. Using GIS and NWS-produced gridded datasets from across the country, the NWS created forecast maps of precipitation, wind, snowfall, and temperatures. These maps are an easy and understandable portrayal of local, regional, and national forecast weather phenomena and are being used throughout the meteorological community.

http://www.weather.gov/btv/

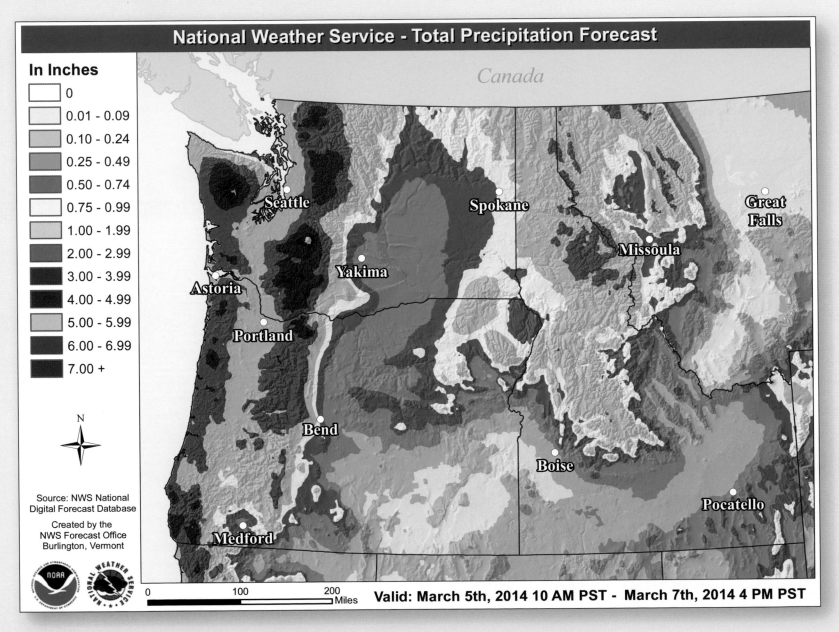

Flash Flood Watch

The National Weather Service (NWS) needed a way to automatically create customized, graphical representations of flash flood warnings for display on its website and for dissemination via social media. Using the ArcGIS ArcPy library, the NWS created this product so forecasters were not burdened with developing this type of graphic by hand. The result has been increased public awareness of flash flood watches; a significant time savings for forecasters; and a consistent, standardized display of information.

http://www.floodsafety.noaa.gov/

31

"It's our job to quantify these (military) events, provide the commander with a clear and concise picture, and then give the commander the understanding that he needs to employ the soldiers throughout this environment."

JASON A. FESER
Geospatial Engineer, US Army

US DEPARTMENT OF DEFENSE

"Let us expand collaboration and integration beyond the intelligence community and Department of Defense to federal, state, and local governments as appropriate."

LETITIA A. LONG
Director Emeritus, National Geospatial-Intelligence Agency

Identifying Potential Sites for the Beneficial Use of Dredged Sediments

Many of the traditional, authorized placement sites for dredged sediments along maintained navigable waterways are at or near capacity. The US Army Corps of Engineers New Orleans District and Engineering Research and Development Center collaborated with stakeholders to identify potential dredged sediment placement sites along the Bayou Segnette Waterway (BSWW) in Louisiana. Landscape analyses identified areas near the BSWW that would benefit from sediment additions. The primary landscapes and conditions recommended for dredged sediment applications include areas of significant erosion (shoreline and interior marsh blow-outs), thin floating marsh adjacent to the BSWW, access canals, and small patches of attached or undetermined marsh types that are below ideal marsh elevation.

 http://www.erdc.usace.army.mil/

LEGEND

- Project boundary
- Access canals
- Floating vegetation
- Eroded land (1956-2008)
- Water

Channel overburden (ft)
- 0.1 - 1
- 1.1 - 2
- 2.1 - 3
- 3.1 - 4
- 4.1 - 5
- 5.1 - 6

Selected range (ft)
- 0.1 - 0.2
- 0.3 - 0.4
- 0.5 - 0.6
- 0.7 - 0.8
- 0.9 - 1
- 1.1 - 1.2
- 1.3 - 1.4
- 1.5 - 1.6

Elevation (ft)
- < -0.7
- -0.7 - -0.6
- -0.5 - -0.4
- -0.3 - -0.2
- -0.1 - 0
- 0.1 - 0.2
- 0.3 - 0.4
- 0.5 - 0.6
- 0.7 - 0.8
- 0.9 - 1
- 1.1 - 1.2
- 1.3 - 1.4
- 1.5 - 1.6
- 1.7 - 1.8
- 1.9 - 2
- 2.1 - 2.2
- 2.3 - 2.4
- 2.5 - 2.6
- 2.7 - 2.8
- 2.9 - 3
- 3.1 - 3.2
- 3.3 - 3.4
- 3.5 - 3.6
- 3.7 - 3.8
- 3.9 - 4
- 4.1 - 4.2
- 4.3 - 4.4
- 4.5 - 4.6
- 4.7 - 4.8

Lake Cataouatche

Lake Salvador

Effects of Mississippi River Bank Crevassing near Fort St. Philip, Louisiana

The US Army Corps of Engineers Mississippi River Valley Division and Engineering Research and Development Center collaborated with the US Geological Survey to learn the effects of constructed crevasses and freshwater diversions on the landscape over time. GIS was used to identify, quantify, and compare the magnitude and sequencing of breaching and impacts associated with the crevassing of the Mississippi River near Fort St. Philip, Louisiana, from 1956 to 2008. Assessments from recent years showed that crevasses may provide long-term benefits in land creation. The study also showed that additional management and/or restoration projects may be important factors that aid crevasses or freshwater diversions in creating land. The results from this study and other research efforts should help guide the planning, implementation, and management of future projects, including the design and construction of sediment diversion projects.

http://www.erdc.usace.army.mil/

Unchanged	2008-1956		2008-1998	1998-1988	1988-1978	1978-1971	1971-1956	Land Change
Land			572	416	738	315	197	Gain (acres)
Water			249	1080	1053	1931	832	Loss (acres)

Grand Coquille Bay

Little Coquille Bay

Fort St. Philip

Mississippi River

Columbia River Treaty

The Columbia River Treaty was signed into law in 1964, and 2014 represented the first opportunity for the United States and Canada to amend or terminate the treaty. Flood-risk management and hydropower generation were the key treaty elements considered in 1964. The Columbia River Treaty Review Program reevaluated these two critical international water management issues while also considering cultural resource, ecosystem, navigation, and climate change elements, among others, that are now factors in the Columbia River Basin. During the review, GIS was used for topographic/bathymetric digital elevation modeling, hydrologic and hydraulic analyses, and climate change modeling. GIS cartography was used at many internal meetings, public open houses, and public listening sessions. The maps shown here were created to give a clear look at the Columbia River Basin and introduce the region and analysis area to a wider audience around the country.

http://www.crt2014-2024review.gov

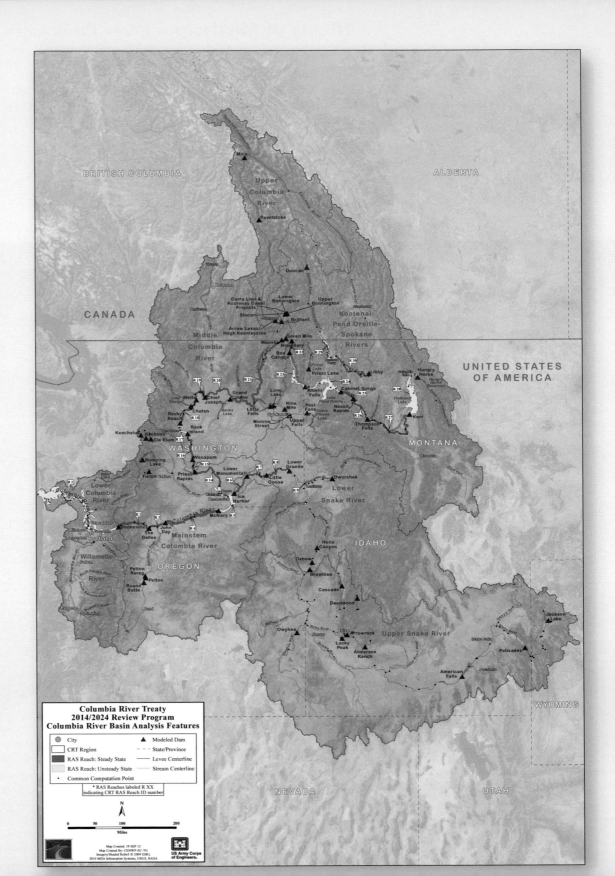

Columbia River Treaty, continued

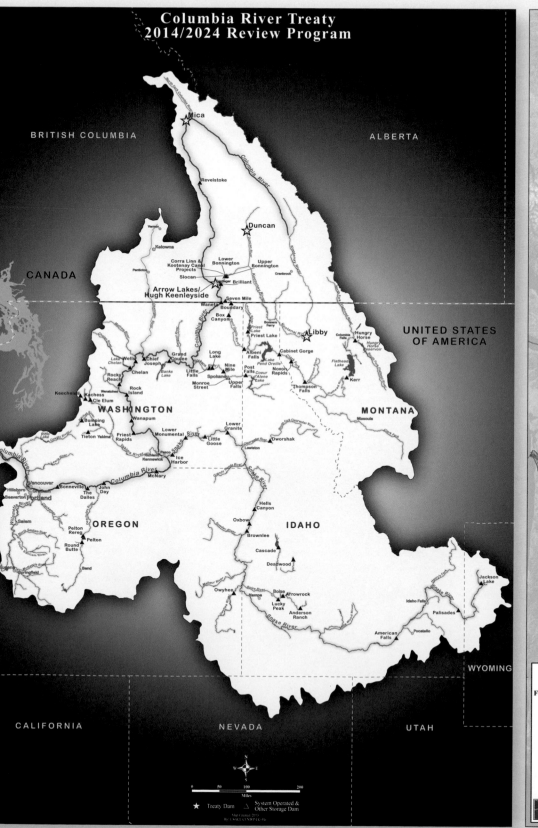

Columbia River Treaty
2014/2024 Review Program

**Columbia River Treaty
2014/2024 Review Program
Columbia River Basin
Flood Risk Assessment Dam & Reservoir System**

- ● City
- ☆ Treaty Dam
- ▲ System Operated & Other Storage Dam
- ☐ CRT Region
- – – State/Province
- — Stream Centerline

Map Created: 02 APR 13
Map Created By: CENWP-EC-TG
Imagery/Shaded Relief © 2009 ESRI,
2010 MDA Information Systems, USGS, NASA

US Army Corps
of Engineers.

Episodic Disturbance on Point Au Fer Island, Louisiana

Misinterpreting the possible causes of localized land loss linked to such episodic events as hurricanes and floods can potentially lead to inappropriate or ineffective ecosystem restorations. The US Army Corps of Engineers New Orleans District and Engineering Research and Development Center worked with Louisiana Coastal Area stakeholders to develop methods and datasets for coastal landscapes to supplement existing information. This study provided a mechanism for testing and using novel methods to expedite the classification of panchromatic aerial photography. These results proved to be critical components of project plan formulation, identifying areas of significant hurricane-induced land loss, and correlations between erosion and soil characteristics. The result was a refinement of proposed and screened restoration measures.

 http://www.erdc.usace.army.mil/

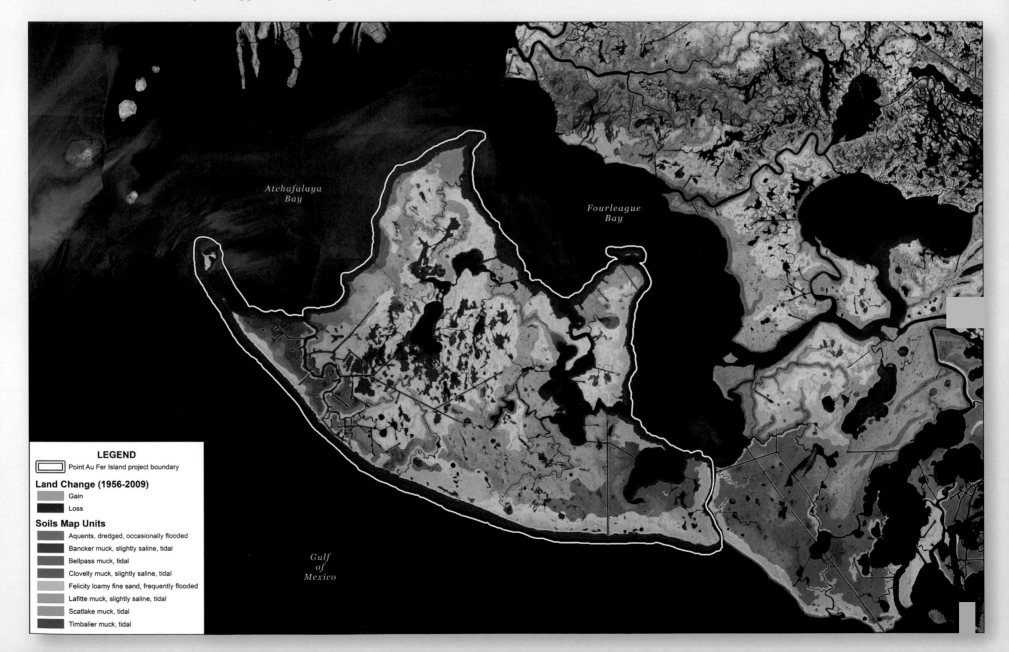

LEGEND

⬜ Point Au Fer Island project boundary

Land Change (1956-2009)
- Gain
- Loss

Soils Map Units
- Aquents, dredged, occasionally flooded
- Bancker muck, slightly saline, tidal
- Bellpass muck, tidal
- Clovelly muck, slightly saline, tidal
- Felicity loamy fine sand, frequently flooded
- Lafitte muck, slightly saline, tidal
- Scatlake muck, tidal
- Timbalier muck, tidal

Atchafalaya Bay

Fourleague Bay

Gulf of Mexico

US Army National Guard Energy Efficiency Map

http://esriurl.com/8473

The US Army Corps of Engineers is responsible for facilities management of more than 165,000 buildings across the nation, many built back in the 1940s. Oblique visible and thermal imagery assists in identifying heat loss and structural conditions for quick correction. The imagery was acquired under optimum environmental conditions using an oblique mapping camera system. This imagery of buildings at Camp Keyes in Augusta, Maine, was used by the Dewberry engineering firm as part of its energy efficiency project for Army National Guard bases to identify thermal anomalies for facility managers.

Navy Energy Consumption Maps

US Navy Secretary Ray Mabus has set the Navy's goal of producing half of its energy from alternative sources by 2020. To help meet that goal, these maps were developed from data contained within the Navy Shore Geospatial Energy Module (NSGEM). NSGEM and derivative products like these deliver summary analysis in a quick and understandable format.

http://www.navy.mil/local/cni/

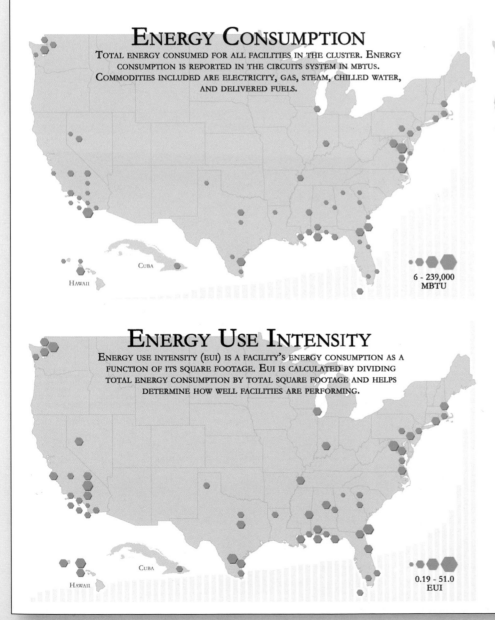

ENERGY CONSUMPTION

TOTAL ENERGY CONSUMED FOR ALL FACILITIES IN THE CLUSTER. ENERGY CONSUMPTION IS REPORTED IN THE CIRCUITS SYSTEM IN MBTUS. COMMODITIES INCLUDED ARE ELECTRICITY, GAS, STEAM, CHILLED WATER, AND DELIVERED FUELS.

CUBA

HAWAII

6 - 239,000
MBTU

FACILITY SQUARE FOOTAGE

TOTAL SQUARE FOOTAGE FOR ALL FACILITIES IN THE CLUSTER. SQUARE FOOTAGE IS MAINTAINED IN THE INFADS REAL PROPERTY SYSTEM. FACILITIES INCLUDED ARE ROOFED STRUCTURES SUITABLE FOR HOUSING PEOPLE, MATERIALS, OR EQUIPMENT.

CUBA

HAWAII

0.19 - 40,666
KSF

ENERGY USE INTENSITY

ENERGY USE INTENSITY (EUI) IS A FACILITY'S ENERGY CONSUMPTION AS A FUNCTION OF ITS SQUARE FOOTAGE. EUI IS CALCULATED BY DIVIDING TOTAL ENERGY CONSUMPTION BY TOTAL SQUARE FOOTAGE AND HELPS DETERMINE HOW WELL FACILITIES ARE PERFORMING.

CUBA

HAWAII

0.19 - 51.0
EUI

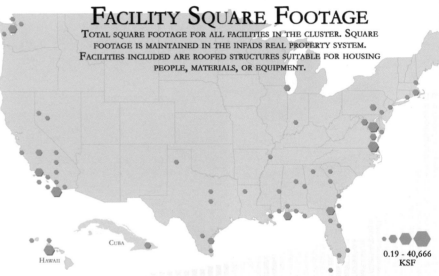

NAVY SHORE GEOSPATIAL ENERGY MODULE

THE PURPOSE OF NSGEM IS TO PROVIDE TOTAL SITUATIONAL AWARENESS OF THE NAVY'S VARIOUS ENERGY PROGRAM ACTIVITIES BY ALLOWING FOR VISUALIZATION OF DISPARATE ENERGY-RELATED DATA SOURCES IN A SINGLE LOCATION. NSGEM CHARTS PROVIDE USERS WITH DATA REGARDING ENERGY CONSUMPTION AND THE PROGRESS OF ENERGY REDUCTION EFFORTS. ADDITIONALLY, THE USE OF GEOSPATIAL DATA ALLOWS FOR THE IMMEDIATE VISUAL ANALYSIS OF THE FACILITY, INSTALLATION, TENANT AND REGIONAL INFORMATION.

NAVY SITES FOR THE CONTINENTAL US, HAWAII, AND CUBA WERE AGGREGATED INTO DATA BINS (CLUSTERS) OF A 100,000 SQ KM GRID.

New Energy Consumption Maps, continued

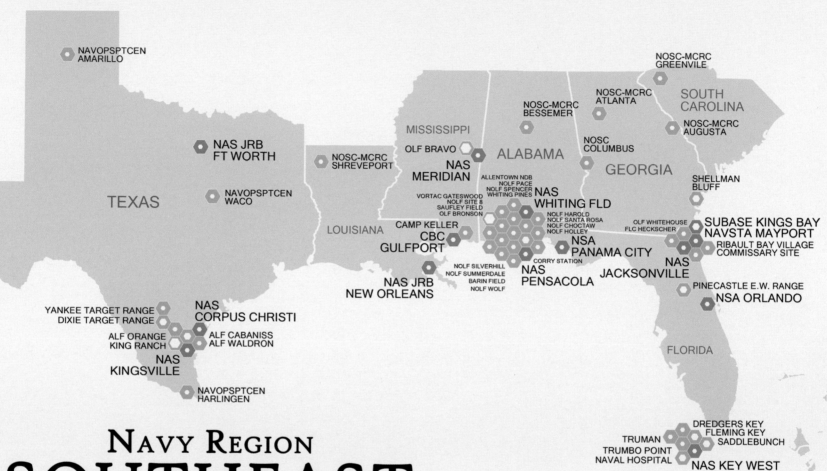

NAVOPSPTCEN
AMARILLO

NAS JRB
FT WORTH

NAVOPSPTCEN
WACO

TEXAS

NOSC-MCRC
SHREVEPORT

LOUISIANA

YANKEE TARGET RANGE
DIXIE TARGET RANGE

NAS
CORPUS CHRISTI

ALF ORANGE
KING RANCH

ALF CABANISS
ALF WALDRON

NAS
KINGSVILLE

NAVOPSPTCEN
HARLINGEN

MISSISSIPPI

OLF BRAVO

NAS
MERIDIAN

VORTAC GATESWOOD
NOLF SITE 8
SAUFLEY FIELD
OLF BRONSON

CAMP KELLER
CBC
GULFPORT

NAS JRB
NEW ORLEANS

NOLF SILVERHILL
NOLF SUMMERDALE
BARIN FIELD
NOLF WOLF

ALLENTOWN NDB
NOLF PACE
NOLF SPENCER
WHITING PINES

ALABAMA

NOSC-MCRC
BESSEMER

NOSC-MCRC
ATLANTA

NOSC
COLUMBUS

GEORGIA

NAS
WHITING FLD

NOLF HAROLD
NOLF SANTA ROSA
NOLF CHOCTAW
NOLF HOLLEY

NSA
PANAMA CITY

CORRY STATION

NAS
PENSACOLA

OLF WHITEHOUSE
FLC HECKSCHER

NAS
JACKSONVILLE

NOSC-MCRC
GREENVILE

SOUTH
CAROLINA

NOSC-MCRC
AUGUSTA

SHELLMAN
BLUFF

SUBASE KINGS BAY
NAVSTA MAYPORT
RIBAULT BAY VILLAGE
COMMISSARY SITE

NAS

PINECASTLE E.W. RANGE
NSA ORLANDO

FLORIDA

DREDGERS KEY
FLEMING KEY
SADDLEBUNCH

TRUMAN
TRUMBO POINT
NAVAL HOSPITAL

NAS KEY WEST

CUBA

NAVSTA
GUANTANAMO BAY

Navy Region
SOUTHEAST
ENERGY CONSUMPTION

NAVY INSTALLATION

NAVY SITE / SPECIAL AREA

LESS THAN 10 MBTU PER 1,000 SQ FT

BETWEEN 10 AND 25 MBTU PER 1,000 SQ FT

MORE THAN 25 MBTU PER 1,000 SQFT

US DEPARTMENT OF EDUCATION

School Attendance Boundary Survey 2013–2014 School Year

No single collection of attendance areas for schools existed on a nationwide level. GIS enabled representatives of the approximately 13,000 regular school districts to upload their boundaries in a digital format, approve boundaries from a previous year's collection effort, or draw boundaries. A publicly accessible mapping system that contains school boundaries for all public schools in this country allows users to examine relationships between schools on a district and national level.

http://nces.ed.gov/surveys/
sdds/ed/ed_express/

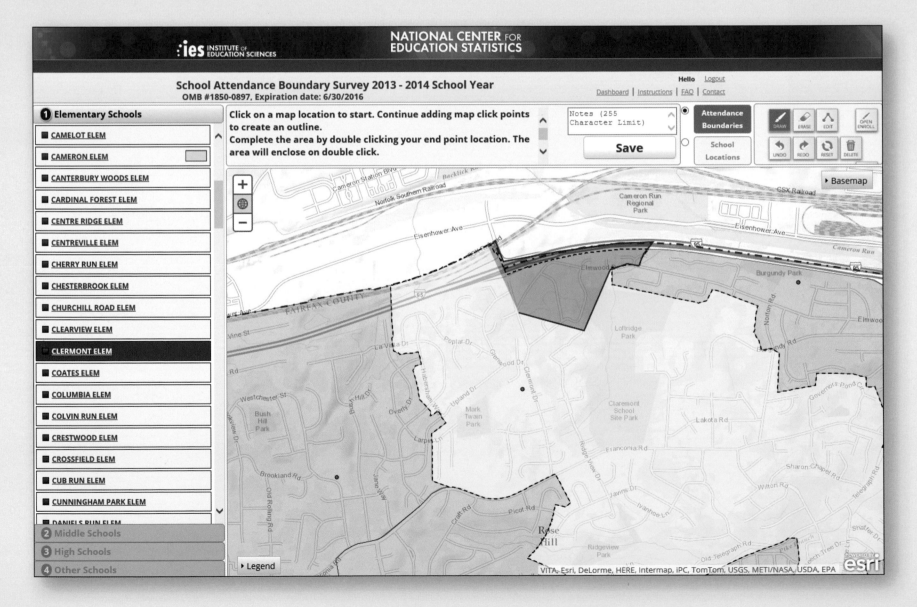

School Attendance Boundary Survey 2013–2014 School Year, continued

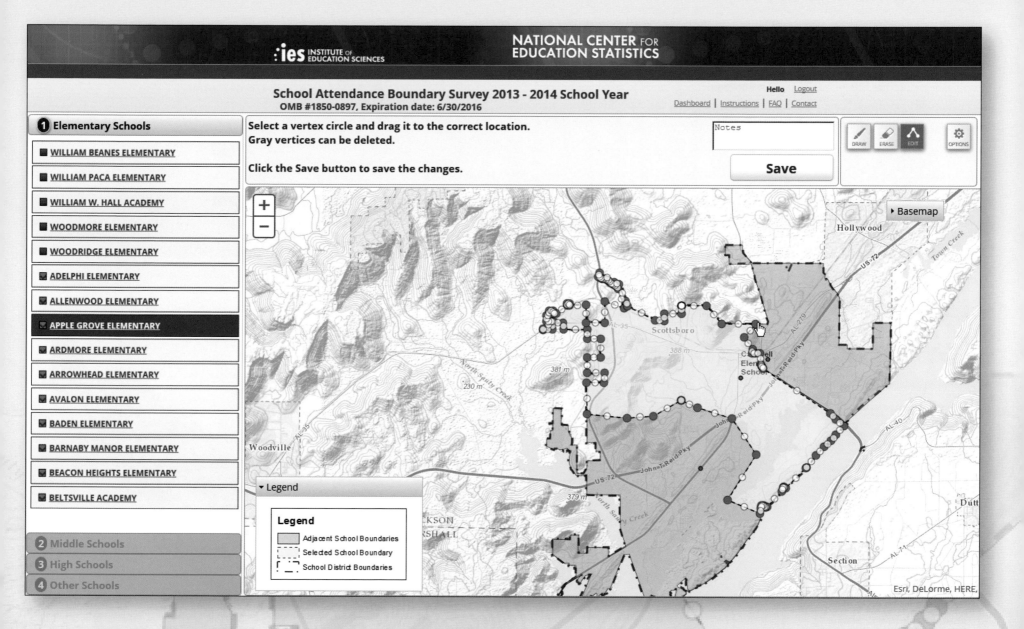

US DEPARTMENT OF EDUCATION

NATIONAL CENTER FOR EDUCATION STATISTICS

Poverty in School-Age Children between 2005 and 2011 for School Districts

Education officials wanted to communicate to the public the problem of school-age children in poverty. This story map helped illustrate the significant increases in poverty in school-age children over the past few years. US Department of Education poverty estimates are used as one of the criteria to allocate federal funds to local education agencies. Also, state and local programs use these estimates for distributing funds and managing school programs.

http://storymaps.blueraster.com/saipe/index.html

Poverty in School Age Children between 2005-2011 for School Districts

Poverty amongst school-age children (5 to 17 living in families) has increased in 26% of US Counties between 2007-2011 according to data from the Small Area Income and Poverty Estimates (SAIPE) program released by the U.S. Census Bureau.

2005　2006　2007　2008　2009　2010　2011

Between 2007 and 2011, the school-age poverty rate (which covers children 5 to 17 living in families) has shown a statistically significant increase in 832 counties, or 26 percent of all counties in the United States, according to data from the Small Area Income and Poverty Estimates (SAIPE) program released by the U.S. Census Bureau. These statistics cover every county and school district in the nation.

"These estimates are sponsored by the U.S. Department of Education and are used as one of the criteria to allocate federal funds to local education agencies," said acting Census Bureau Director Thomas Mesenbourg. "In addition, state and local programs use these estimates for distributing funds and managing school programs."

Push the play button, or explore a particular year, to see how Poverty among school-age children has increased or decreased within the country's Unified and Elementary School Districts between 2005-2011.

Read the news release

School-age Poverty Rate
- Greater than 30%
- 20 - 30%
- 10 - 20%
- Less than 10%

POWERED BY esri

US DEPARTMENT OF HEALTH AND HUMAN SERVICES

Protecting Shellfish Growing Areas

The US Food and Drug Administration conducted a technical information exchange with Korea Shellfish Sanitation Program staff. The project showed how to visualize actual and potential pollution sources affecting shellfish growing areas, based on the flow rate of the source. GIS was used to interpret GPS data and determined and/or verified the size of prohibited zones established around wastewater treatment plant discharges in proximity to shellfish growing areas. This method delivered precise and accurate local-level information, bolstered a science-based food safety policy/program, and also helped optimize the data-collector/investigator work plan and resource allocation.

http://esriurl.com/8475

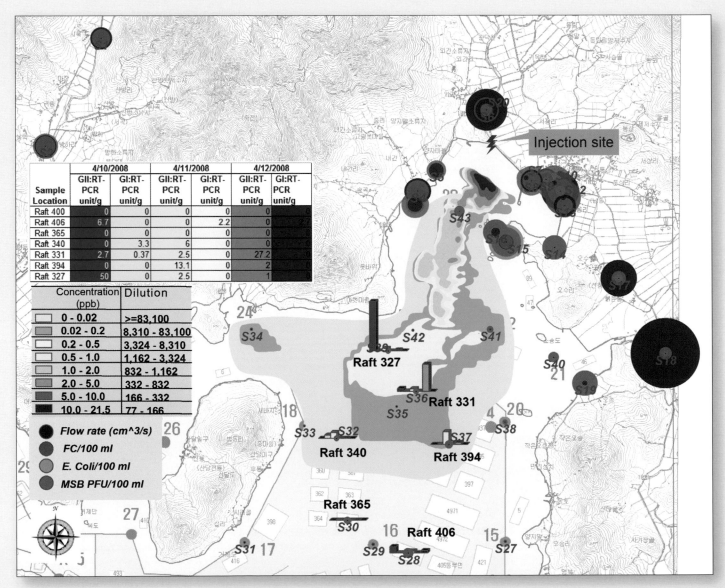

Sample Location	4/10/2008		4/11/2008		4/12/2008	
	GII:RT-PCR unit/g	GI:RT-PCR unit/g	GII:RT-PCR unit/g	GI:RT-PCR unit/g	GII:RT-PCR unit/g	GI:RT-PCR unit/g
Raft 400	0	0	0	0	0	0
Raft 406	6.7	0	0	2.2	0	2.7
Raft 365	0	0	0	0	0	0
Raft 340	0	3.3	6	0	0	0
Raft 331	2.7	0.37	2.5	0	27.2	0
Raft 394	0	0	13.1	0	2	0
Raft 327	50	0	2.5	0	1	0

Concentration (ppb)	Dilution
0 - 0.02	>=83,100
0.02 - 0.2	8,310 - 83,100
0.2 - 0.5	3,324 - 8,310
0.5 - 1.0	1,162 - 3,324
1.0 - 2.0	832 - 1,162
2.0 - 5.0	332 - 832
5.0 - 10.0	166 - 332
10.0 - 21.5	77 - 166

Flow rate (cm^3/s)
FC/100 ml
E. Coli/100 ml
MSB PFU/100 ml

Injection site

Protecting Shellfish Growing Areas, continued

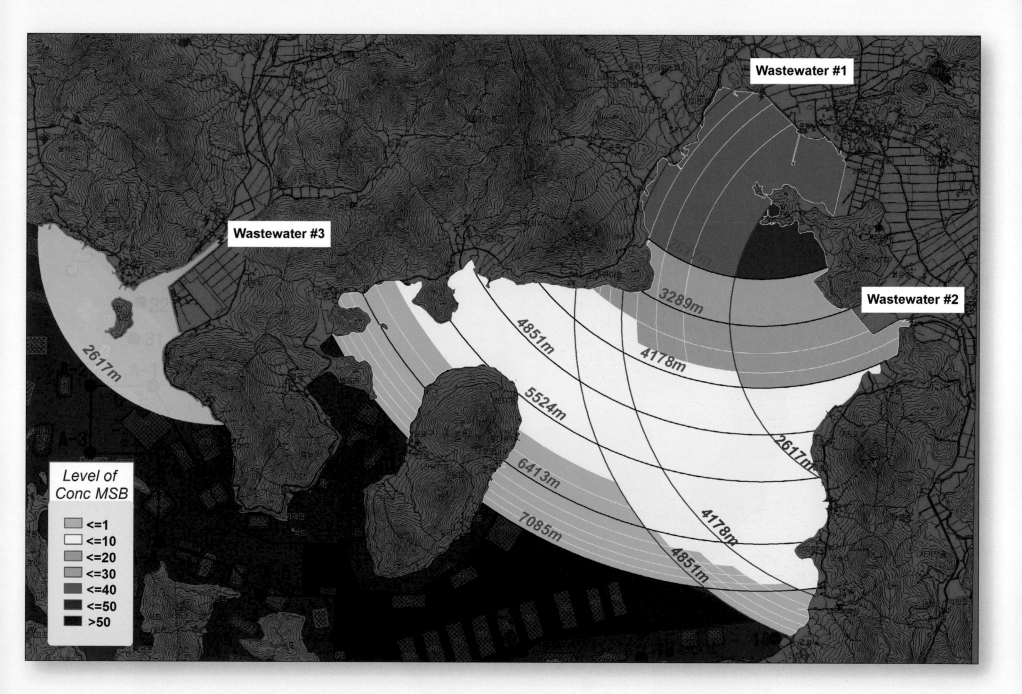

Age-Adjusted Death Rates for United States, 2006–2010

The National Cancer Institute provides dynamic views of cancer and related statistics for prioritizing cancer-control efforts in the United States. Age-adjusted incidence and death rates per 100,000 are computed, along with statistics for demographic and screening and risk factor variables, and stored in a database. Using GIS to connect to a database with over a million records allows the website user to create dynamic maps with many variables, colors, and groups.

 http://www.cancer.gov/

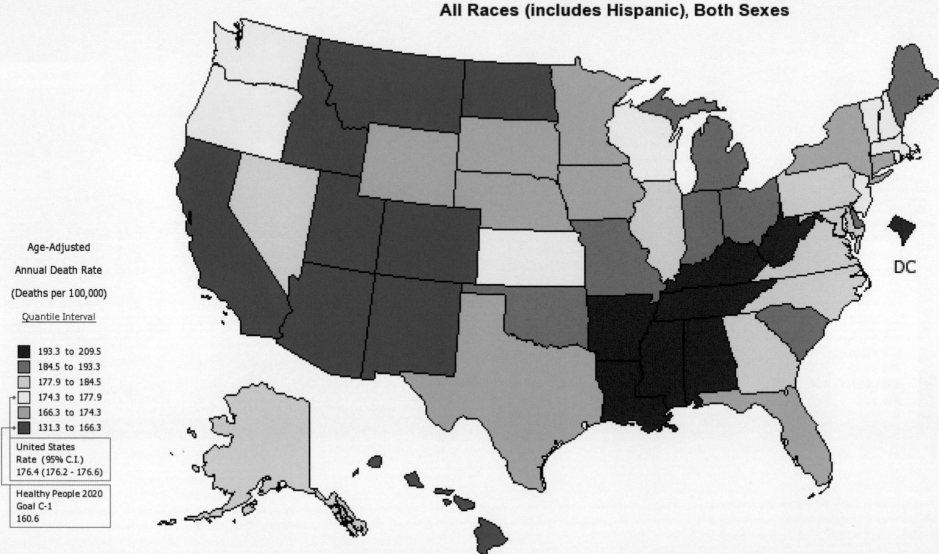

Age-Adjusted Death Rates for United States, 2006 - 2010
All Cancer Sites
All Races (includes Hispanic), Both Sexes

Age-Adjusted
Annual Death Rate
(Deaths per 100,000)

Quantile Interval

■	193.3 to 209.5
■	184.5 to 193.3
■	177.9 to 184.5
■	174.3 to 177.9
■	166.3 to 174.3
■	131.3 to 166.3

United States
Rate (95% C.I.)
176.4 (176.2 - 176.6)

Healthy People 2020
Goal C-1
160.6

TOXMAP

http://www.nlm.nih.gov/pubs/factsheets/toxmap.html

TOXMAP was deployed in 2004 as an interactive, public-facing GIS for searching toxic release data from the US Environmental Protection Agency. But map technology evolved and the old TOXMAP appeared dated. In 2014, a next-generation TOXMAP based on ArcGIS for Server and Adobe Flex technology was released. These maps show the Toxics Release Inventory, Superfund, Canadian toxic release facilities, and US commercial nuclear power plants. The new TOXMAP provides better interactive capabilities with a GIS look and feel. It also offers improved place finding and new and improved data layers, such as the US Census.

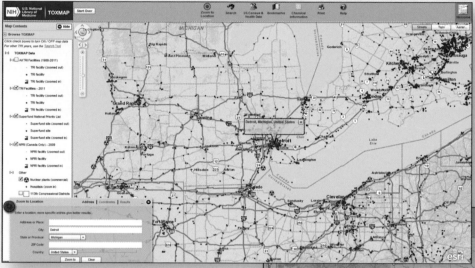

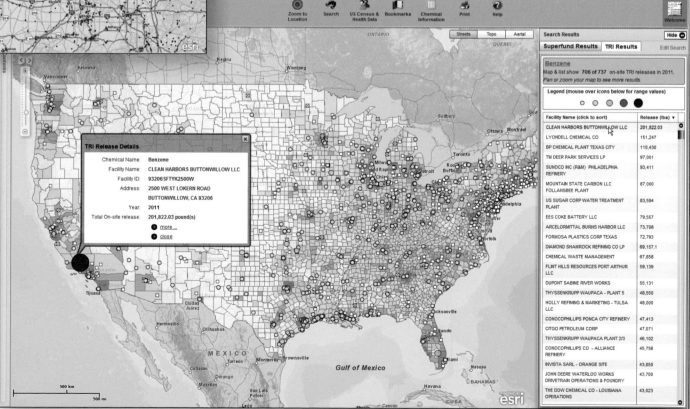

US DEPARTMENT OF HOMELAND SECURITY

Driving the Nation 2011–2013

The US Coast Guard Navigation Center operates the Nationwide Differential GPS, which assesses signal quality and coverage. Prior to GIS, signal measurements were drawn using a graphics painting program. Signal measurements are now georeferenced using ArcGIS software and GPS signal quality analysis has improved.

Service managers and system engineers make more informed decisions in changing operational parameters and equipment to meet the mandated service availability requirements.

http://esriurl.com/8474

Driving the Nation: 2011 - 2013
Validating the Nationwide Differential GPS (NDGPS) Coverage

Isabela

Pacific Region Map for US Coast Guard

The US Coast Guard wanted a map featuring the Pacific Ocean floor with shaded relief bathymetry showing fracture zones, basins, ridges, and sea mounts. Although primarily a decorative piece, the map produced by Maps.com for the Coast Guard conveys information that would not normally be included on a standard geopolitical map. Data provided by the National Oceanic and Atmospheric Administration and the Coast Guard was used to create the map using marine colors and shading to show the seafloor features with clear, detailed labeling.

http://www.uscg.mil/pacarea/

National Flood Insurance Program

Maps are important to the National Flood Insurance Program (NFIP) because they clarify the program, its requirements, and its effects, and also help to communicate the risk of residing within a flood-prone area. Paper maps proved inefficient, so NFIP modernized its mapping program using ArcGIS Online. The National Flood Hazard Layer online map, updated daily, makes it easy to find a location and extract information from the features, enhancing risk communication and public

access to information. The map showing NFIP Flood Insurance Policy concentrations throughout the nation gives a perspective on where people are buying down the risk through the purchase of flood insurance. Many flood insurance policies have premiums that are lower than their risk would indicate due to grandfathering and other subsidies. Recent NFIP flood insurance reform legislation has sought to remedy this. The map showing the concentration of NFIP insurance policies where

subsidies may be removed has helped explain the effects of this legislation to policy makers as well as the general public.

The National Flood Hazard Layer
(Flood Zones where FIRMs are Modernized)

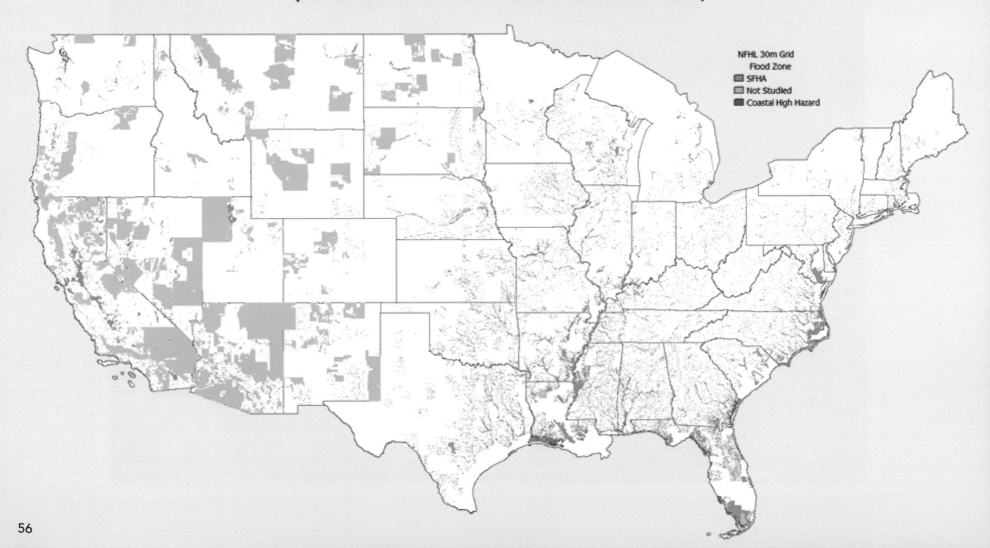

NFHL 30m Grid
Flood Zone
▢ SFHA
▢ Not Studied
▣ Coastal High Hazard

National Flood Insurance Program, continued

http://www.fema.gov/
national-flood-insurance-program

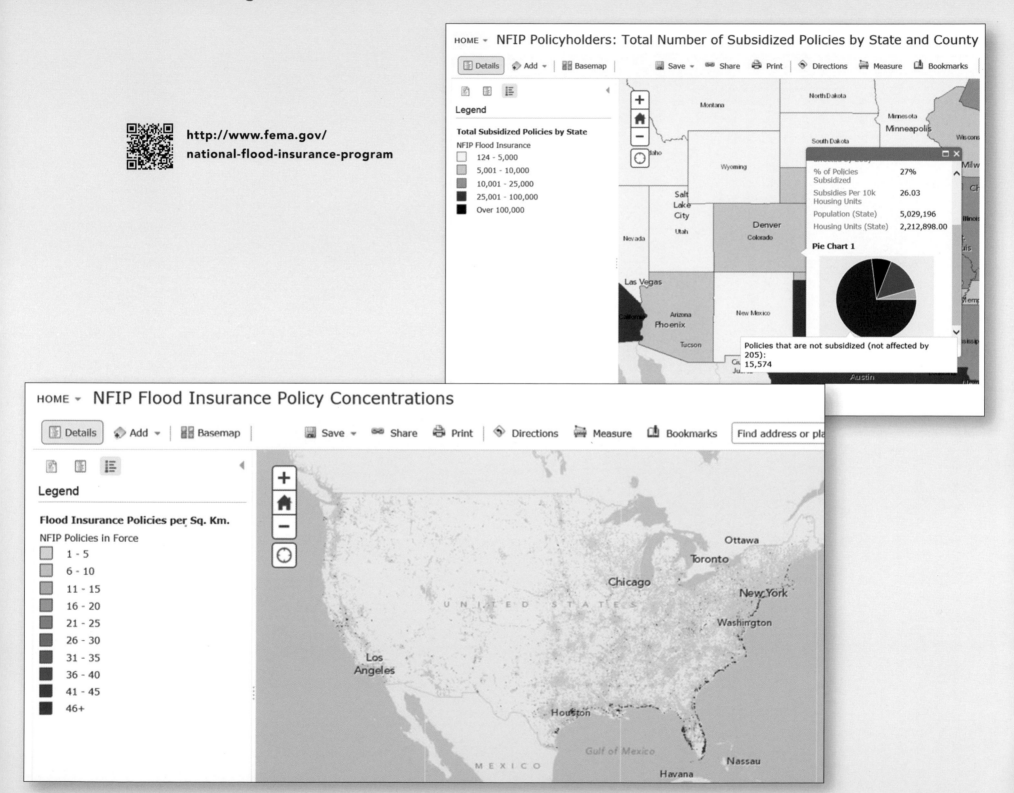

US DEPARTMENT
OF THE INTERIOR

Offshore North Carolina Visualization Study

While planning for offshore wind power in North Carolina, the Bureau of Ocean Energy Management sought accurate depictions of offshore wind turbines to help evaluate potential visual impacts. This map shows areas with completed visual models of potential wind farm locations in the Atlantic. Stakeholders see what different turbine groups, sizes, and distances from shore might look like in different light and atmospheric conditions. This map helps the bureau and its stakeholders quickly understand the potential visual impacts of wind development.

http://esriurl.com/8476

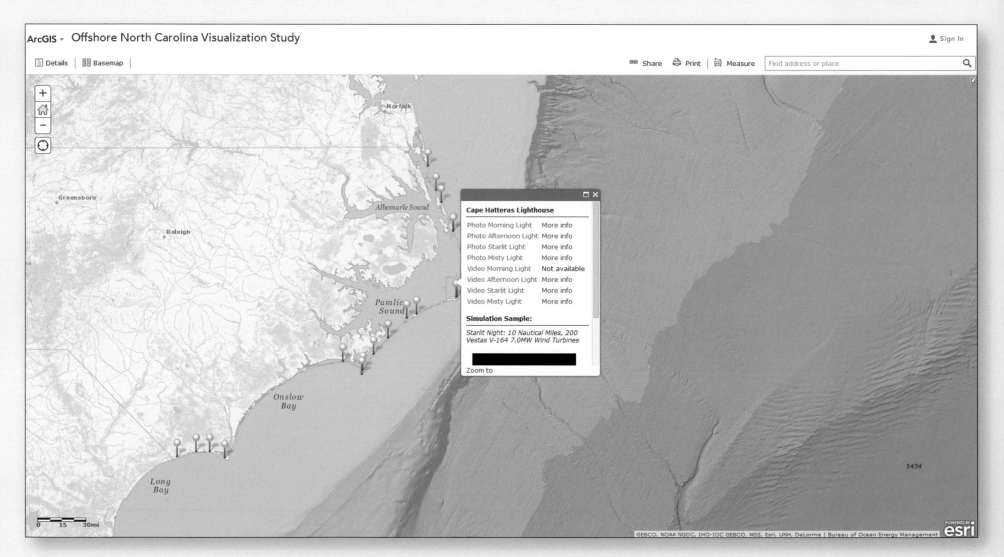

Offshore North Carolina Visualization Study, continued

Simulation
006 Cape Hatteras Lighthouse
Late Afternoon
Siemens SWT-3.6-107
10 nm

GENERAL INFORMATION

Base Photograph
Photo Name: HLA_0001-UV
Date: April 14, 2012
Time: 3:50 PM
GPS Coordinates[1]: lat 35.250515°, long -75.528816°
Viewpoint Elevation: 172'

Sun and Weather
Sun Angle/Azimuth: 258°
Sun Elevation: 33°
Lighting Angle: Side lit
Weather Conditions: Sunny
Visibility[2]: 10 mi
Wave Height: 2 - 4'
Period: 4 sec.

Camera
Camera Make/Model: Nikon D7000
Sensor Dimensions: 23.6 mm X 15.6 mm
Lens Make/Model: Nikkor DX AF-S 35 mm
Lens Focal Length: 35 mm
35 mm Equivalent Focal Length: 52.5 mm
Horizontal and Vertical Angles of View:
     37.3° wide and 25.3° high
Camera Height: 1.5 m (5')
Camera Azimuth[3]: 200°

Wind Turbine Information
Number: 200
Make and Model: Siemens SWT-3.6-107
Height/Dimensions:
  Support Structure/Monopile Ht.: 13 m (43')
  Hub Ht. (above Monopile): 80 m (262')
  Rotor Diameter: 107 m (351')
  Total Height to Tip of Blade: 147 m (481')
  Service Platform: A bldg. 50'H X 100'W X 200' L
    elevated 50' above the water

CONTEXT MAP

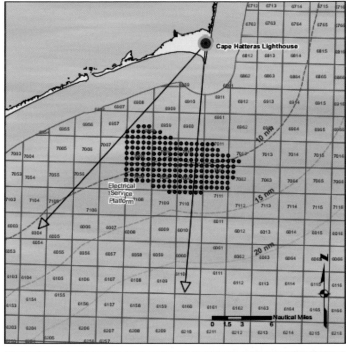

VIEWING INSTRUCTIONS

The simulation is properly printed on an 11" X 17" sheet at actual size. If viewed on a computer monitor, use the highest screen resolution. The simulated image is at the proper perspective when viewed at 23.5" from the eye, or at a distance of approximately twice the image height.

NOTES

- Turbine heights were adjusted to the image's horizon line, which was slightly curved due to camera lens distortion.
- The image was taken with a UV filter.
- Refraction Coefficient[4] (k) = .075

PANORAMA

37.3° x 25.3°

Simulation location within the panorama view (190° X 60°)
from the Cape Hatteras Lighthouse site

T. J. Boyle Associates
landscape architects • planning consultants

Showcasing Success Stories in MarineCadastre.gov

The Bureau of Ocean Energy Management and the National Oceanic and Atmospheric Administration provide authoritative data to meet the needs of the offshore energy and marine planning communities. This joint effort needed an effective way to document the breadth of applications for such a wealth of data and tools. Points on this map link to user success stories from across the country to show how different organizations use MarineCadastre.gov data and products to meet their specific needs. Users can quickly see the broad range of uses that MarineCadastre.gov supports, including evaluating impacts of offshore energy on navigation safety, researching the effects of noise from large commercial vessels on marine mammals, and creating maps of proposed wave energy projects.

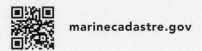

marinecadastre.gov

Tiber Reservoir Map

http://esriurl.com/8477

The public had problems locating campgrounds, recreation areas, amenities, roads, and structures around Tiber Reservoir in north central Montana. The Bureau of Reclamation posted this information map with roads, campgrounds, recreation areas, and structures that accurately shows distance, reservoir topography, contact information, and other structures. The map enhanced visitor satisfaction and reduced public complaints.

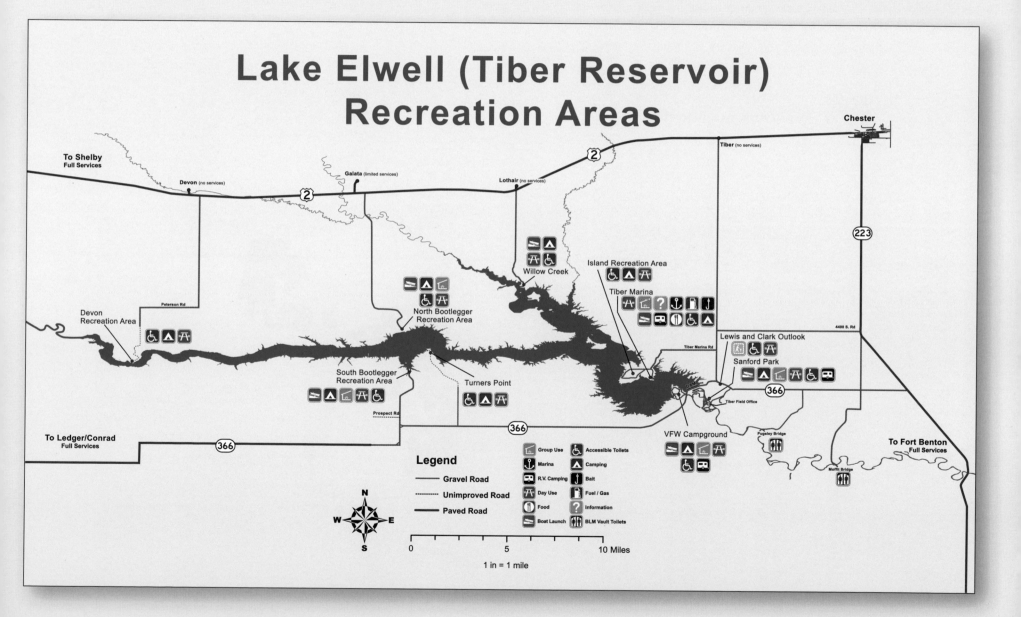

National Wildlife Refuge System

The US Fish and Wildlife Service needed a standard method to depict existing and newly established National Wildlife Refuges on a US map. Established in 1903, the National Wildlife Refuge system has grown to include more than 560 refuges, 38 wetland management districts, and other protected areas encompassing 150 million acres of land and water from the Caribbean to the remote Pacific. The Division of Realty's cartographers created a standard map template and update process using the strengths of the ArcGIS platform and Adobe Illustrator software based on the Fish and Wildlife Service's spatial data holdings. Whenever a new refuge is created or a refuge name changes, the map is easily updated and made available online.

http://www.fws.gov/refuges/

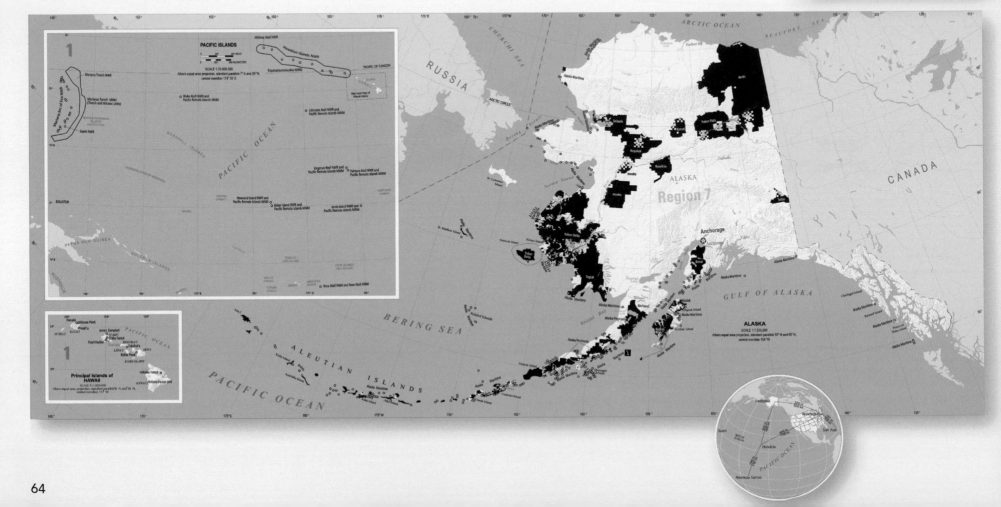

National Wildlife Refuge System, continued

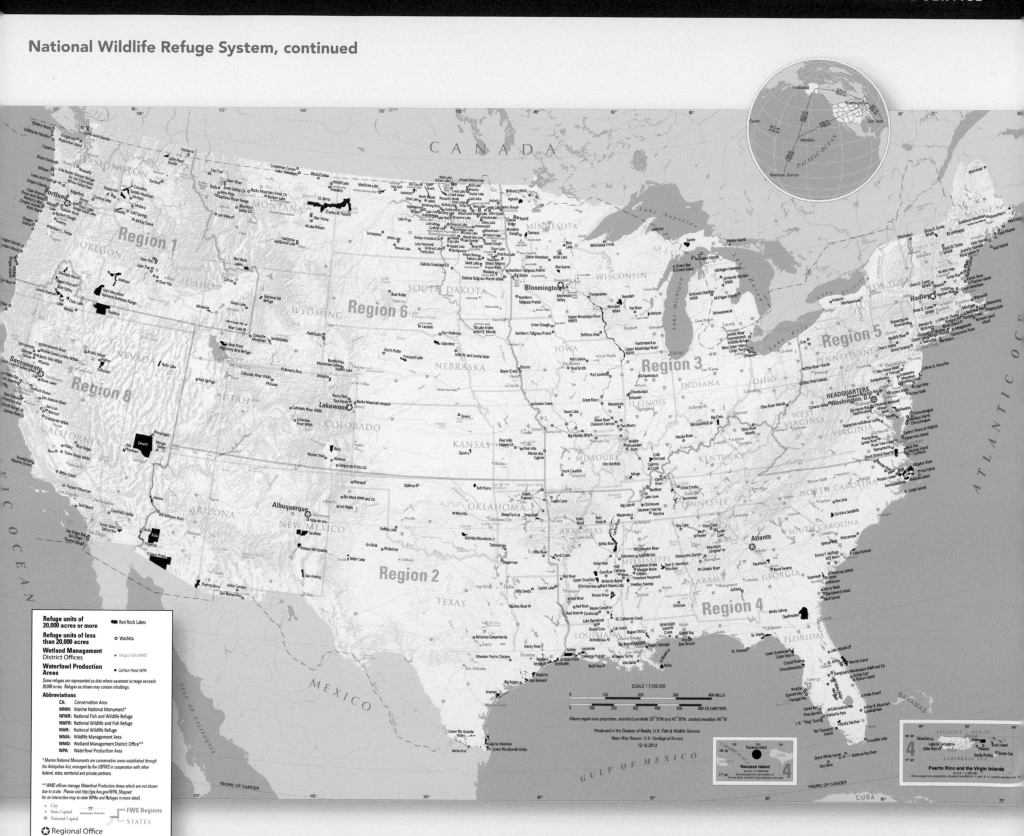

Refuge units of 20,000 acres or more — ▬ Red Rock Lakes

Refuge units of less than 20,000 acres — ○ Washita

Wetland Management District Offices — ▲ Fergus Falls WMD

Waterfowl Production Areas — ● Carlton Pond WPA

Some refuges are represented as dots where easement acreage exceeds 20,000 acres. Refuges as shown may contain inholdings.

Abbreviations

CA: Conservation Area
MNM: Marine National Monument*
NFWR: National Fish and Wildlife Refuge
NWFR: National Wildlife and Fish Refuge
NWR: National Wildlife Refuge
WMA: Wildlife Management Area
WMD: Wetland Management District Office**
WPA: Waterfowl Production Area

* Marine National Monuments are conservation areas established through the Antiquities Act, managed by the USFWS in cooperation with other federal, state, territorial and private partners.

** WMD offices manage Waterfowl Production Areas which are not shown due to scale. Please visit http://gis.fws.gov/WPA_Mapper/ for an interactive map to view WPAs and Refuges in more detail.

● City
◉ State Capital
⊛ National Capital
✪ Regional Office

FWS Regions
STATES

Download a copy at: www.fws.gov/refuges/maps/

SCALE 1:7,500,000

Albers equal area projection, standard parallels 29°30'N and 45°30'N, central meridian 96°W

Produced in the Division of Realty, U.S. Fish & Wildlife Service
Base Map Source: U.S. Geological Survey
12-16-2013

Navassa Island

Puerto Rico and the Virgin Islands

US DEPARTMENT OF VETERANS AFFAIRS

VA Enrollee Population by County

The Veterans Health Administration Policy and Planning Office wanted to communicate the value that GIS contributes to its overall mission. It published a map on the office's website to give veterans and the public a general idea of where the veteran enrollee population resides. GIS helps visualize choice and explain the spatial dimension of health care as it relates to optimizing healthcare access and resources. Visual data-driven analysis often impacts policies and quality of care for veterans.

**http://www.va.gov/vetdata/
Veteran_Population.asp**

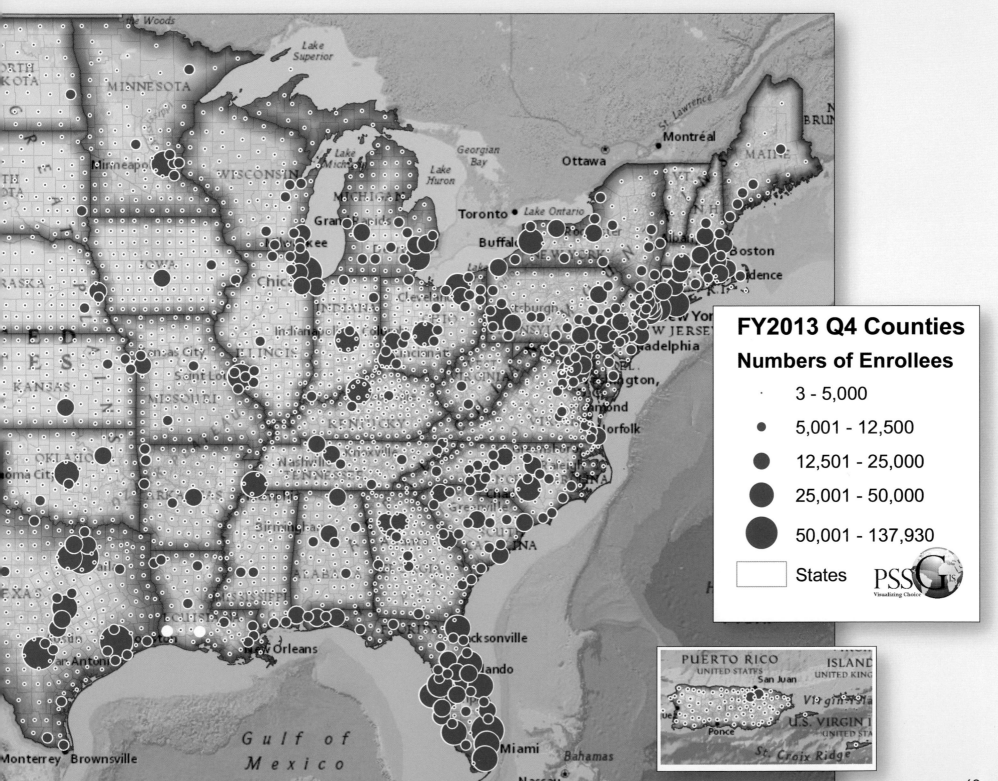

FY2013 Q4 Counties

Numbers of Enrollees

· 3 - 5,000

• 5,001 - 12,500

● 12,501 - 25,000

● 25,001 - 50,000

● 50,001 - 137,930

☐ States

PSSG
Visualizing Choice

INDEPENDENT FEDERAL AGENCIES AND TRIBAL GOVERNMENTS

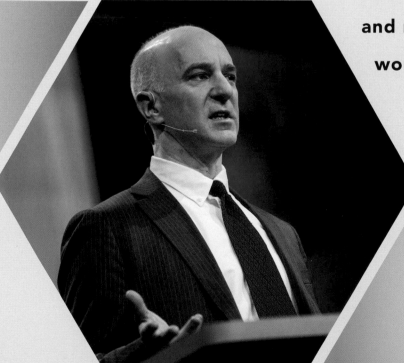

"The digital strategy and the executive order for open data at their heart are about opening the hood of government and making it possible to see the inner workings."

HARVEY SIMON

Geographic Information Officer, US Environmental Protection Agency

How's My Waterway?

"How's My Waterway?" is the Environmental Protection Agency's mobile web app and website that helps people learn the condition of their local waterways using a smartphone, tablet, or desktop computer. This map shows the Potomac River in Washington, DC. The online service makes science-based water quality information accessible and understandable for everyone.

http://watersgeo.epa.gov/mywaterway/map.html

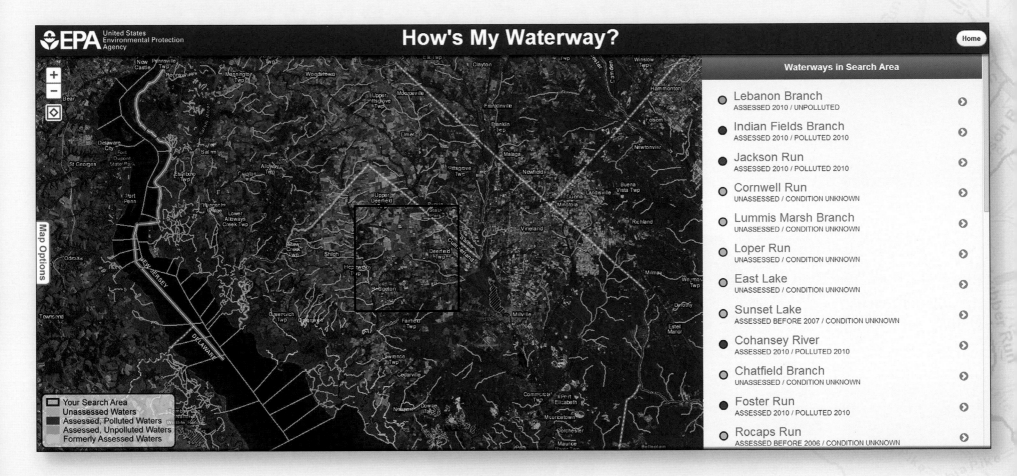

How's My Waterway?, continued

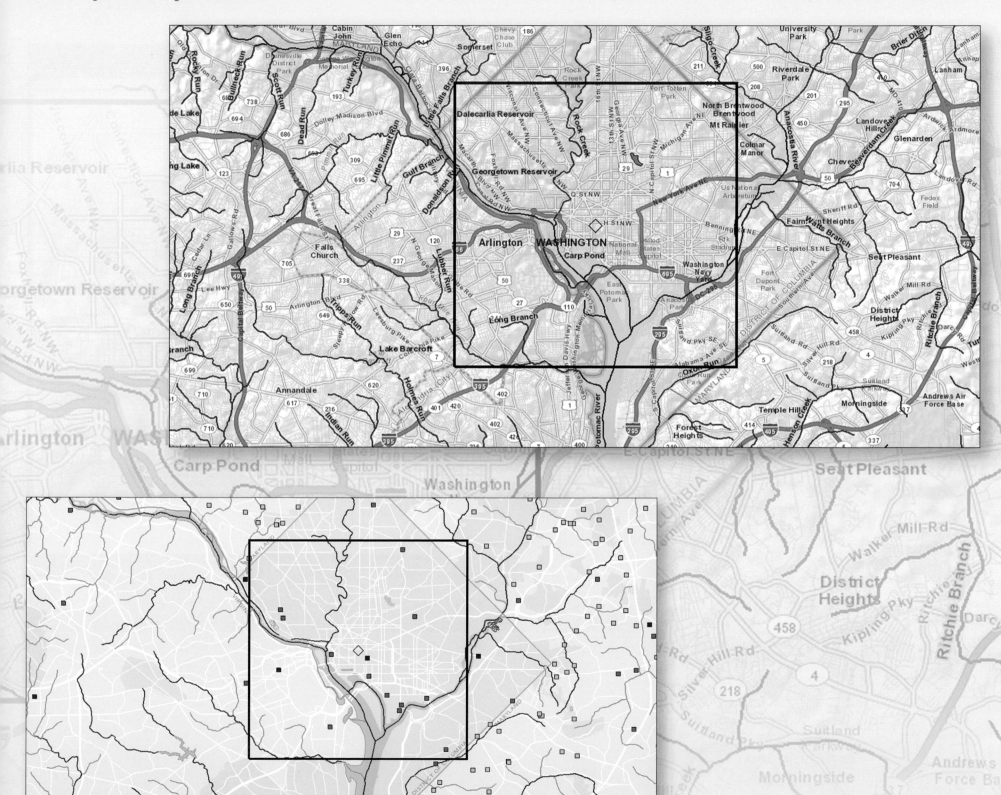

Facility Registry System on the Exchange Network

The Exchange Network is a partnership among states, tribes, territories, and the Environmental Protection Agency for the exchange of environmental information efficiently and securely over the Internet. The Facility Registry System was the first project to exchange data regarding facilities, sites, or places of environmental interest that are subject to regulation. By displaying the data quality geographically, the application provides an easily visible alert to data stewards to records in the database that need to be improved. Data stewards can more quickly and effectively ensure records in the database are as up to date as possible.

http://www.epa.gov/enviro/
html/fii/Exchange_Network.htm

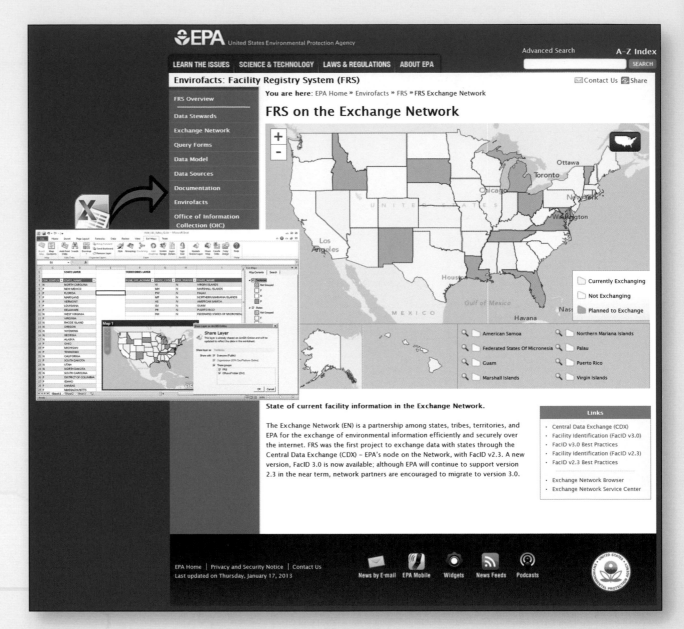

Consolidation Opportunities

 This QR Code links to the program that manages the GSA's internal application referenced on this page.

The US General Services Agency (GSA), landlord for the federal civilian government, needs to help its customer agencies make smart decisions that both minimize costs and align with federal sustainability goals, such as walkability and transit access. ArcGIS analysis allows the GSA to visualize data and understand the relevant spatial relationships so decision making is more efficient and effective. Account managers can quickly compare agencies' current locations with available vacant space in nearby federally owned and leased buildings, and prioritize more sustainable locations.

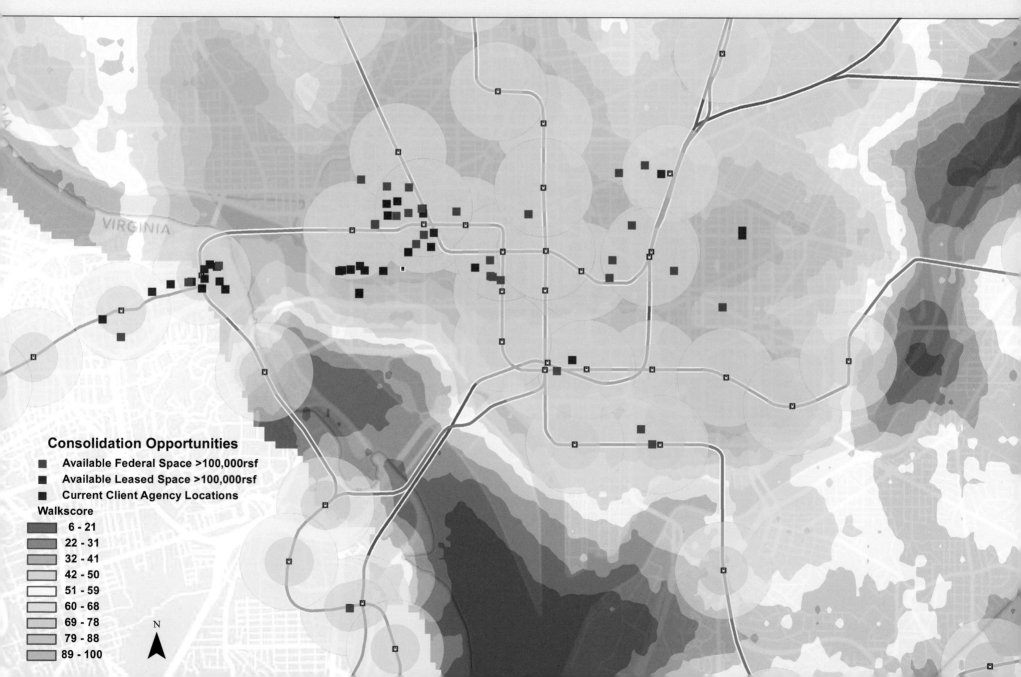

Consolidation Opportunities

- Available Federal Space >100,000rsf
- Available Leased Space >100,000rsf
- Current Client Agency Locations

Walkscore

- 6 - 21
- 22 - 31
- 32 - 41
- 42 - 50
- 51 - 59
- 60 - 68
- 69 - 78
- 79 - 88
- 89 - 100

VIRGINIA

N

GSA Mapping Portal

The US General Services Administration (GSA) has over 12,000 employees around the country, most of whom have little or no GIS training and experience but who can benefit from access to GIS tools and spatial analysis. The GSA's ArcGIS Online Mapping Portal makes GIS more accessible to its workforce. The portal is a one-stop shop for access to GIS tools to share sample maps and analyses across the agency, helping new users employ GIS to inform and enhance their own business practices.

 This QR Code links to the program that manages the GSA's internal application referenced on this page.

HOME   GALLERY   MAP   GROUPS   MY CONTENT   MY ORGANIZATION

General Services Administration
Mapping Portal

GSA Owned and Leased Buildings by Proportional

GSA Community Planning and Sustainability

Active Hurricane Service

(Beta) New Construction and Repair and Alteration (R&A)

Welcome to ArcGIS Online for GSA, a mapping portal that makes it easy to create, analyze and share maps. Now, you can create interactive maps and share them with the rest of the agency. Click the "Map" link above to start building your own maps!

Esri.com | ArcGIS Marketplace | Help | Terms of Use | Privacy | Contact Esri | Report Abuse

GSA's Multi-Asset Planning Tool

The US General Services Administration (GSA) needed a way for its employees with limited GIS experience to use the technology in their work. GSA's Multi-Asset Planning (MAP) tool makes spatial analysis and mapmaking accessible to experts and new users alike. Users can filter building, lease, and client characteristics, do basic distance and area measurements, and identify buildings of interest within a given area. Leasing specialists around the country use MAP to identify clients in expiring high-cost leases who could move into cheaper vacant federal space nearby. The tool also shows client agencies where their space may be at risk of flooding and identifies opportunities to target more transit-accessible sites.

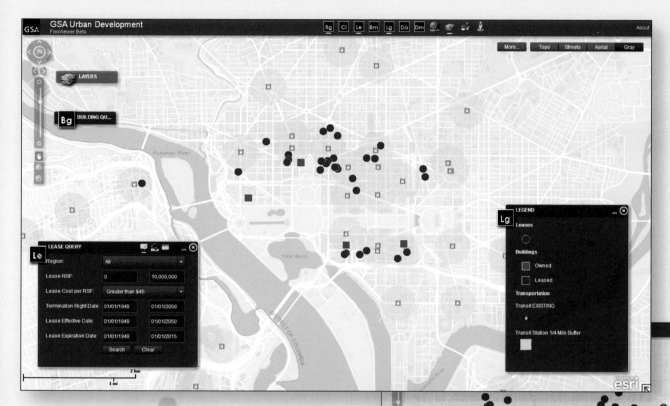

This **QR Code** links to the program that manages the GSA's internal application referenced on this page.

GSA Major Construction Project Map

The US General Services Administration (GSA) wanted to promote President Obama's open government and open data initiatives by sharing data about GSA construction and major renovation and alteration projects. This map displays projects for the 2014 fiscal year that were funded as part of the Consolidated Appropriations Act of 2014 and 2015 fiscal year projects that were included in the president's budget request. Using the map, local stakeholders can quickly identify current and potential future projects in their communities and the associated budget outlays.

 This QR Code links to the program that manages the GSA's internal application referenced on this page.

Community Planning and Sustainability Planning Map

 This QR Code links to the program that manages the GSA's internal application referenced on this page.

The US General Services Administration (GSA) wanted to share information about its community planning and sustainability work with the public. This story map shows the GSA's high-performance green building projects around the country and its community engagement through the public use of federal space and planning outreach and partnership meetings. Story map viewers have access to data about Leadership in Energy and Environmental Design (LEED) certifications, Energy Star scores and innovative green technology, and other sustainability investments in GSA buildings.

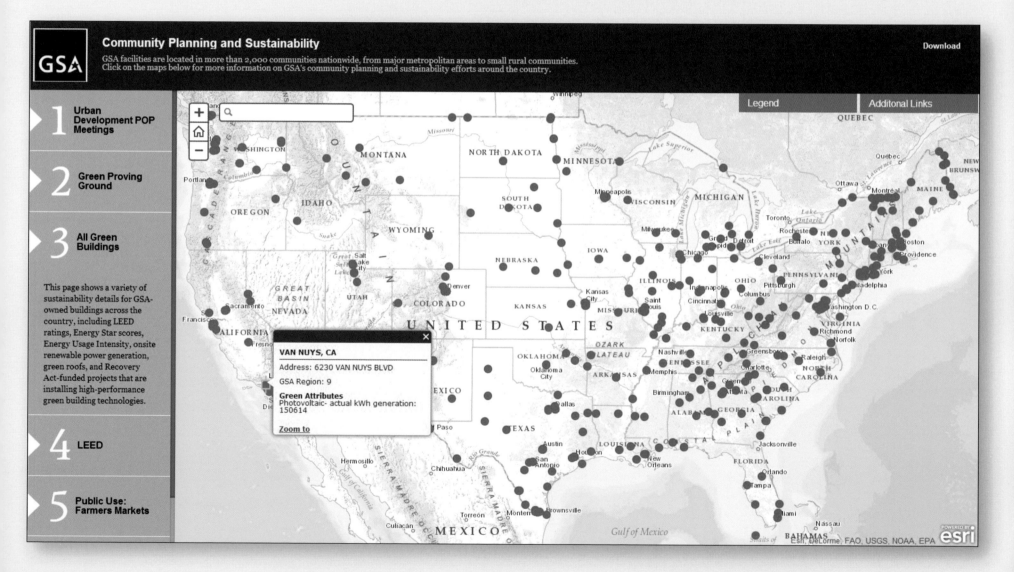

GSA

Community Planning and Sustainability

GSA facilities are located in more than 2,000 communities nationwide, from major metropolitan areas to small rural communities. Click on the maps below for more information on GSA's community planning and sustainability efforts around the country.

Download

1 Urban Development POP Meetings

2 Green Proving Ground

3 All Green Buildings

This page shows a variety of sustainability details for GSA-owned buildings across the country, including LEED ratings, Energy Star scores, Energy Usage Intensity, onsite renewable power generation, green roofs, and Recovery Act-funded projects that are installing high-performance green building technologies.

4 LEED

5 Public Use: Farmers Markets

Legend | Additonal Links

VAN NUYS, CA
Address: 6230 VAN NUYS BLVD
GSA Region: 9
Green Attributes
Photovoltaic- actual kWh generation: 150614
Zoom to

Esri, DeLorme, FAO, USGS, NOAA, EPA

POWERED BY esri

Grex Leasing Tool

The US General Services Administration (GSA) leasing specialists needed to analyze how new lease offers compared to prevailing rental rates in a given real estate submarket. This JavaScript viewer allows leasing specialists to draw in a delineated area boundary and determine which real estate submarkets it falls within. This quick and simple overlay analysis lets leasing specialists compare the offers they receive with average rental rates for a given submarket so they can make more informed selection decisions. The viewer also shows how much of the delineated area falls within the 100-year floodplain and the seismic requirements for buildings within the delineated area.

 This QR Code links to the program that manages the GSA's internal application referenced on this page.

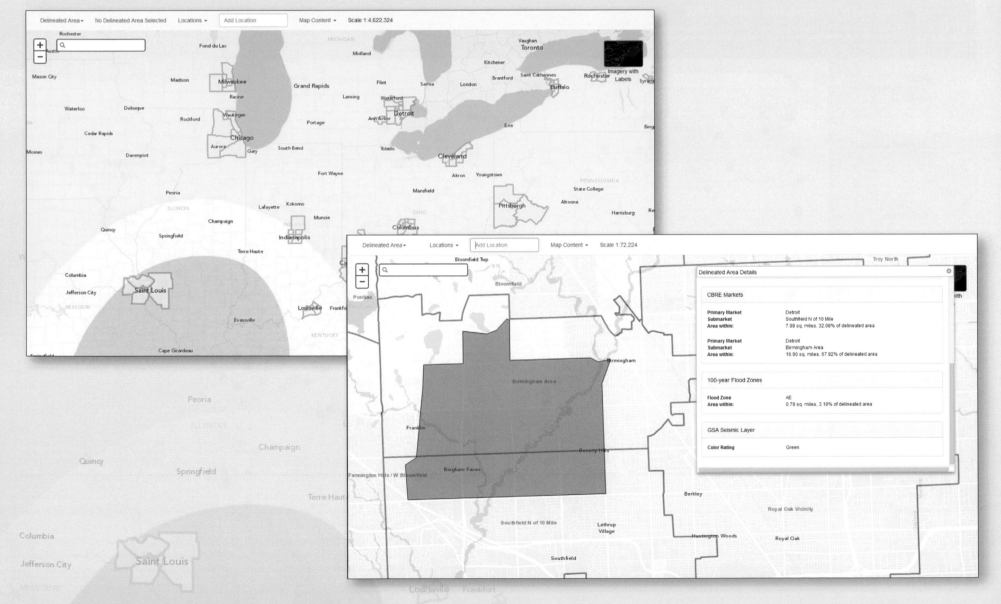

Honolulu Transit Maps

 This QR Code links to the program that manages the GSA's internal application referenced on this page.

The US General Services Administration (GSA) held meetings with local planners and stakeholders in Honolulu to identify opportunities to support the city's future light rail station and planned Transit-Oriented Development (TOD). GSA's urban development staff used the ArcGIS Online Mapping Portal to show the existing and future transit system, land use, TOD locations, and expiring federal leases outside of a walkable distance from new light rail stops. Local and state planners, federal agencies, and congressional staff were able to identify opportunities for consolidation and relocation of federal leases to more sustainable TOD sites.

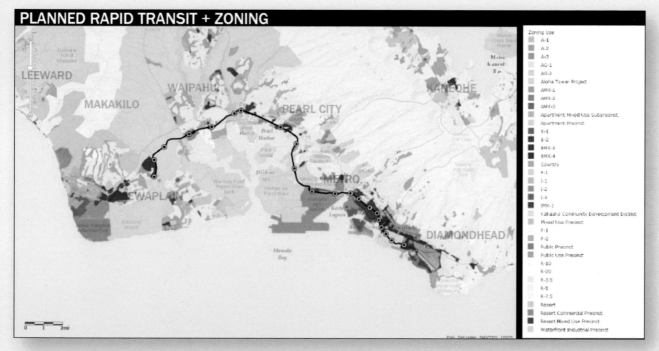

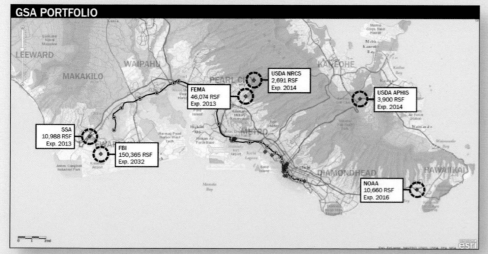

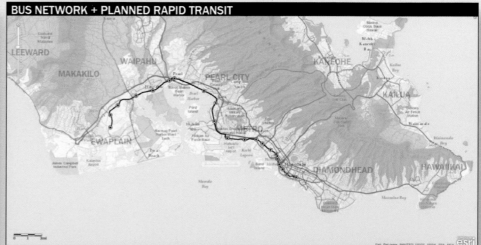

Voyager Card Fuel Transactions 2012–2013

The US Postal Service Office of Investigations wanted to determine if trucking companies contracted to move mail from postal facilities across the United States were fraudulently using fuel cards issued to them. Geospatial tools established the optimal route a truck should take for each trip. All fuel card transactions were then geocoded and mapped on top of these routes. Users were instantly able to identify transaction points which did not fall along the designated routes. These outliers were flagged as suspicious transactions to be investigated further. A total of 6,000 fuel transactions were flagged with an estimated $2.9 million in potentially illegal charges.

The application referenced on this page is internal. This QR Code links to the US Postal Service Office of the Inspector General home page.

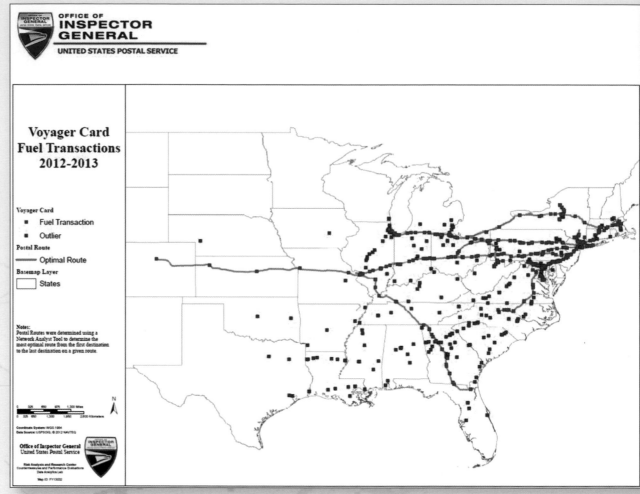

Office of Inspector General Relocation Map

Employees working at an Office of Inspector General (OIG) facility that would soon be closing for consolidation purposes each needed new office locations. GIS assisted in the relocation of OIG employees. Employees' home addresses were mapped along with OIG facility locations that had space available. A drive-time analysis from the closing office to an employee's home address determined how long each employee currently commuted to work. This data was then used to establish new work locations for employees, keeping in mind their current commute time while finding a new facility location for them that fit those criteria.

The application referenced on this page is internal. This QR Code links to the US Postal Service Office of the Inspector General home page.

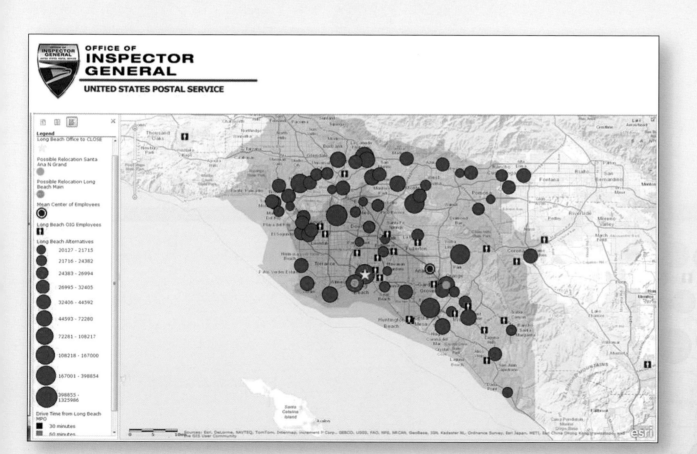

Smithsonian Institution Explorer Maps

The Smithsonian Institution wanted an intranet-based system for viewing and extracting its facilities' geospatial and tabular data. SI Explorer provides an intuitive, user-friendly, self-help environment for the Smithsonian's technical and nontechnical staff. SI Explorer's connection to a facilities management database provides access to information about existing conditions. Shown here are maps of the Smithsonian Castle and adjacent art museums, National Air and Space Museum, National Museum of the American Indian, D.W. Reynolds Center, and National Zoological Park. The easy web-based availability of information has eliminated hundreds of hours of data searching across Smithsonian offices prior to the start of new projects for planning and maintenance or design and construction. SI Explorer's greater purpose is to support improved planning and management decision making via the visual display of key space data.

 http://www.si.edu/

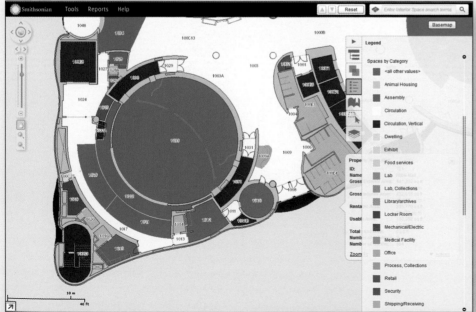

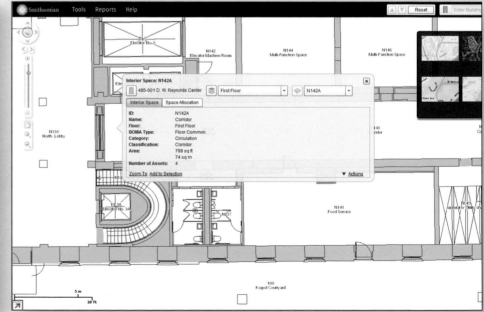

Smithsonian Institution Explorer Maps, continued

Land Use Public Comment Map

The Confederated Tribes (Walla Walla, Umatilla, and Cayuse) of the Umatilla Indian Reservation in the Pacific Northwest needed feedback on proposed reservation development plans. This interactive map allows tribal community members to review proposed projects, submit public comments, and review comments from other members of the community to see what is being said about proposed land-use activities. Sites are listed in the left navigation pane. Different proposed development options are listed below each site. Each concept is clickable, and will display a pop-up window with additional details, including design draws. The interactive map encourages community participation in the planning and development process.

http://ctuir.org/

Explanation of River Mile Discrepancies in Rice, Washington

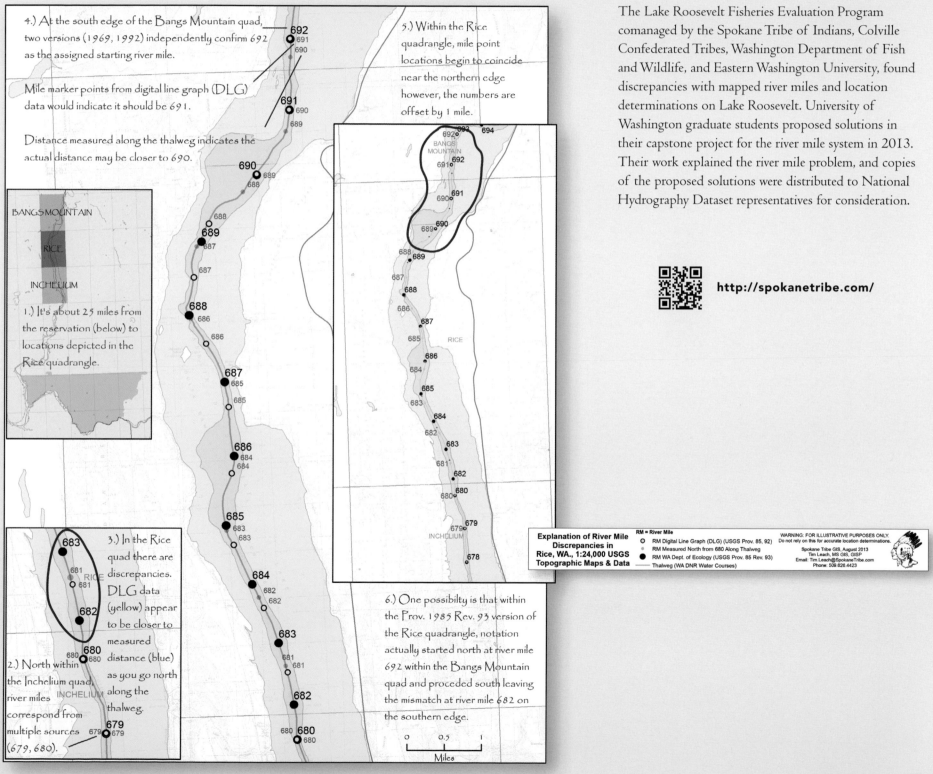

4.) At the south edge of the Bangs Mountain quad, two versions (1969, 1992) independently confirm 692 as the assigned starting river mile.

Mile marker points from digital line graph (DLG) data would indicate it should be 691.

Distance measured along the thalweg indicates the actual distance may be closer to 690.

BANGS MOUNTAIN

RICE

INCHELIUM

1.) It's about 25 miles from the reservation (below) to locations depicted in the Rice quadrangle.

3.) In the Rice quad there are discrepancies. DLG data (yellow) appear to be closer to measured distance (blue) as you go north along the thalweg.

2.) North within the Inchelium quad, river miles correspond from multiple sources (679, 680).

5.) Within the Rice quadrangle, mile point locations begin to coincide near the northern edge however, the numbers are offset by 1 mile.

BANGS MOUNTAIN

RICE

INCHELIUM

6.) One possibilty is that within the Prov. 1985 Rev. 93 version of the Rice quadrangle, notation actually started north at river mile 692 within the Bangs Mountain quad and proceded south leaving the mismatch at river mile 682 on the southern edge.

The Lake Roosevelt Fisheries Evaluation Program comanaged by the Spokane Tribe of Indians, Colville Confederated Tribes, Washington Department of Fish and Wildlife, and Eastern Washington University, found discrepancies with mapped river miles and location determinations on Lake Roosevelt. University of Washington graduate students proposed solutions in their capstone project for the river mile system in 2013. Their work explained the river mile problem, and copies of the proposed solutions were distributed to National Hydrography Dataset representatives for consideration.

http://spokanetribe.com/

Explanation of River Mile Discrepancies in Rice, WA., 1:24,000 USGS Topographic Maps & Data

RM = River Mile
○ RM Digital Line Graph (DLG) (USGS Prov. 85, 92)
· RM Measured North from 680 Along Thalweg
● RM WA Dept. of Ecology (USGS Prov. 85 Rev. 93)
— Thalweg (WA DNR Water Courses)

WARNING: FOR ILLUSTRATIVE PURPOSES ONLY.
Do not rely on this for accurate location determinations.

Spokane Tribe GIS, August 2013
Tim Leach, MS GIS, GISP
Email: Tim.Leach@SpokaneTribe.com
Phone: 509.626.4423

0   0.5   1
Miles

CREDITS

GovLoop and Esri.

US DEPARTMENT OF AGRICULTURE

Early Access to National Agriculture Imagery Program Digital Files, created by US Department of Agriculture (USDA) Farm Service Agency (FSA) Aerial Photography Field Office; data from USDA FSA, Esri, DeLorme, HERE, TomTom, Intermap, Increment P Corp., General Bathymetric Chart of the Oceans, US Geological Survey, Food and Agriculture Organization of the United Nations, National Park Service, Natural Resources Canada, GeoBase, Institut Géographique National, Kadaster Netherlands, Ordnance Survey, Esri Japan, METI (Japan Ministry of Economy, Trade, and Industry), Esri China (Hong Kong), swisstopo, and the GIS User Community.

National Agriculture Imagery Program 2013 Feedback, created by US Department of Agriculture (USDA) Farm Service Agency (FSA) Aerial Photography Field Office; data from USDA FSA, Esri, DeLorme, HERE, TomTom, Intermap, Increment P Corp., General Bathymetric Chart of the Oceans, US Geological Survey, Food and Agriculture Organization of the United Nations, National Park Service, Natural Resources Canada, GeoBase, Institut Géographique National, Kadaster Netherlands, Ordnance Survey, Esri Japan, METI (Japan Ministry of Economy, Trade, and Industry), Esri China (Hong Kong), swisstopo, and the GIS User Community.

Bighorn National Forest Wildfires 1900–2013, created by and data from US Department of Agriculture Forest Service.

Forest Health Maps, created by and data from US Department of Agriculture Forest Service Forest Health Technology Enterprise Team.

2012 Census of Agriculture Atlas, created by US Department of Agriculture National Agricultural Statistics Service (NASS); data from NASS Cropland Data Layers for years 2008–2012, Moderate Resolution Imaging Spectroradiometer.

Ports on Mexico-US Border and Tracking the Drug Route, created by US Department of Agriculture (USDA) Office of Homeland Security and Emergency Coordination (OHSEC); data from USDA OHSEC, Esri, DeLorme, HERE, US Geological Survey, Intermap, Increment P Corp., Natural Resources Canada, Esri Japan, METI (Japan Ministry of Economy, Trade, and Industry), Esri China (Hong Kong), Esri (Thailand), TomTom.

2013 Inauguration Support, created by US Department of Agriculture (USDA) Office of Homeland Security and Emergency Coordination (OHSEC); data from USDA OHSEC, Esri, DeLorme, HERE, TomTom, Intermap, Increment P Corp., General Bathymetric Chart of the Oceans, US Geological Survey, Food and Agriculture Organization of the United Nations, National Park Service, Natural Resources Canada, GeoBase, Institut Géographique National, Kadaster Netherlands, Ordnance Survey, Esri Japan, METI (Japan Ministry of Economy, Trade, and Industry), Esri China (Hong Kong), swisstopo, and the GIS User Community.

US DEPARTMENT OF COMMERCE

Comparing Metro and Micro Area Population Change between 2002–2003 and 2012–2013, created by US Census Bureau; data from US Census Bureau, Esri, HERE, DeLorme, MapmyIndia, ©OpenStreetMap contributors.

County Population Growth between 2012 and 2013 and the Primary Source of Population Change, created by US Census Bureau; data from US Census Bureau, Esri, HERE, DeLorme, MapmyIndia, ©OpenStreetMap contributors.

Metropolitan and Micropolitan Statistical Areas of the United States and Puerto Rico, created by and data from US Census Bureau.

New England City and Town Areas, created by and data from US Census Bureau.

PUMA Reference Map: San Bernardino County and Redlands and Yucaipa Cities, created by US Census Bureau; data from US Census Bureau's TIGER database.

SAIPE Derived Poverty Surface: Poverty Ratio for Related Children Ages 5 to 17, created by US Census Bureau; data from US Census Bureau's TIGER database and Small Area Income and Poverty Estimates (SAIPE) Program.

High-Resolution Coastal Change Analysis Program, created by and data from National Oceanic and Atmospheric Administration's Coastal Services Center.

Washington State Spatial Prioritization Data Viewer, created by and data from National Oceanic and Atmospheric Administration National Centers for Coastal Ocean Services, Center for Coastal Monitoring and Assessment, Biogeography Branch.

Mapping Coral Reefs, created by and data from National Oceanic and Atmospheric Administration National Centers for Coastal Ocean Services, Center for Coastal Monitoring and Assessment, Biogeography Branch.

Estimated Changes New Datums Will Create, created by and data from National Oceanic and Atmospheric Administration National Geodetic Survey.

Hillside Visualizations of National Ocean Service Bathymetric Attributed Grids, created by National Oceanic and Atmospheric Administration (NOAA) National Geophysical Data Center (NGDC); data from NOAA NGDC, US National Ocean Service Hydrographic Database, US Geological Survey, Monterey Bay Aquarium Research Institute, US Army Corps of Engineers, Shuttle Radar Topography Mission, International Bathymetric Chart of the Caribbean Sea and the Gulf of Mexico, National Aeronautics and Space Administration.

CREDITS

Natural Hazards Viewer, created by National Oceanic and Atmospheric Administration (NOAA) National Geophysical Data Center (NGDA); data from NOAA NGDA, General Bathymetric Chart of the Oceans (GEBCO_08 Grid, version 20100927), ETOPO1 Global Relief Model, Esri, DeLorme, HERE, MapmyIndia, ©OpenStreetMap contributors, DigitalGlobe, Earthstar Geographics, National Geographic, National Aeronautics and Space Administration, Geonames.org, CNES/Airbus DS, GeoEye, I-Cubed, US Department of Agriculture Farm Service Agency, US Geological Survey, AEX, Getmapping, Aerogrid, ESA, METI (Japan Ministry of Economy, Trade, and Industry), Natural Resources Canada, Institut Géographique National, Instituto Geográfico Português, swisstopo, and the GIS User Community.

Hurricane Sandy Reconstruction, created by National Oceanic and Atmospheric Administration (NOAA) National Ocean Service (NOS); data from NOAA NOS, NOAA Coastal Services Center, Federal Emergency Management Agency, US Army Corps of Engineers, US Global Change Research Program, New York City Office of Long-Term Planning, Esri, DeLorme, NAVTEQ, TomTom, Digital Globe, GeoEye, I-Cubed, US Department of Agriculture, US Geological Survey, AEX, Getmapping, Aerogrid, Institut Géographique National, Instituto Geográfico Português, swisstopo, and the GIS User Community.

National Weather Service Maps (Highs, Lows, Current Temperatures), created by National Oceanic and Atmospheric Administration (NOAA) National Weather Service (NWS); data from NOAA NWS National Digital Forecast Database.

National Weather Service Total Precipitation Forecast, created by National Oceanic and Atmospheric Administration (NOAA) National Weather Service (NWS); data from NOAA NWS National Digital Forecast Database.

Flash Flood Watch, created by National Oceanic and Atmospheric Administration (NOAA) National Weather Service (NWS); data from NOAA NWS, Esri, DigitalGlobe, GeoEye, I-Cubed, US Department of Agriculture, US Geological Survey, AEX, Getmapping, Aerogrid, Institut Géographique National, Instituto Geográfico Português, and the GIS User Community.

US DEPARTMENT OF DEFENSE

Identifying Potential Sites for the Beneficial Use of Dredged Sediment, created by US Army Engineer Research and Development Center; data from US Army Corps of Engineers, Louisiana State University, US Geological Survey.

Effects of Mississippi River Bank Crevassing near Fort St. Philip, Louisiana, created by US Army Engineer Research and Development Center; data from Army Map Service, National Aeronautics and Space Administration, Louisiana Oil Spill Coordinator's Office, US Army Corps of Engineers, US Geological Survey.

Episodic Disturbance on Point Au Fer Island, Louisiana, created by US Army Engineer Research and Development Center; data from US Geological Survey, Natural Resources Conservation Service.

Columbia River Treaty, created by US Army Corps of Engineers (USACE); data from USACE, MDA Information Systems, US Geological Survey, National Aeronautics and Space Administration, TomTom, StreetMap North America, World Shaded Relief, Bureau of Transportation Statistics, National Transportation Atlas Databases, Esri, US Environmental Protection Agency, US Department of Agriculture.

US National Guard Efficiency Map, created by and data from US Army Corp of Engineers, Dewberry, and ©2014 Icaros, Inc.

Navy Energy Consumption Maps, created by Geographic Information Services, Inc.; data from Navy Shore Geospatial Energy Module.

US DEPARTMENT OF EDUCATION

School Attendance Boundary Survey 2013–2014 School Year, created by Blue Raster and Institute for Educational Sciences (IES); data from IES National Center for Educational Statistics, US Census Bureau, Esri, DeLorme, HERE, VITA, Intermap, iPC, TomTom, US Geological Survey, Ministry of Economy, Trade, and Industry (METI) of Japan/National Aeronautics and Space Administration, US Department of Agriculture, US Environmental Protection Agency.

Poverty in School-Age Children between 2005 and 2011 for School Districts, created by Blue Raster and US Census Bureau; data from US Census Bureau Small Income and Poverty Estimates program, Esri, DeLorme, HERE, MapmyIndia, ©OpenStreetMap contributors.

US DEPARTMENT OF HEALTH AND HUMAN SERVICES

Protecting Shellfish Growing Areas, created by and data from Food and Drug Administration Center for Food Safety and Applied Nutrition, Office of Food Safety, and Korea Shellfish Sanitation Program.

Age-Adjusted Death Rates for the United States 2006–2012, created by National Cancer Institute (NCI); data from NCI, IMS, Inc., National Vital Statistics Section, Centers for Disease Control and Prevention, US Census, state cancer registries.

TOXMAP, created by US National Library of Medicine (NLM); data from NLM, US Environmental Protection Agency, Hazardous Substances Database, Agency for Toxic Substances and Disease Registry, National Atlas of the United States of America, National Cancer Institute, National Weather Service, Esri, GeoGratis.

US DEPARTMENT OF HOMELAND SECURITY

Driving the Nation 2011-2013, created by and data from US Coast Guard Navigation Center.

Pacific Region Map for US Coast Guard, created by 2014 Maps.com; data from 2014 Maps.com, US Coast Guard.

National Flood Insurance Program, created by and data from Federal Emergency Management Agency.

For more information on GIS in the federal government, visit esri.com/federal.

US DEPARTMENT OF THE INTERIOR

Offshore North Carolina Visualization Study, created by Bureau of Ocean Energy Management (BOEM); data from BOEM, Esri Oceans Basemap, General Bathymetric Chart of the Oceans, National Oceanic and Atmospheric Administration, National Geographic, DeLorme, International Hydrographic Organization-Intergovernmental Oceanographic Commission, National Geodetic Survey, University of New Hampshire. Not to be used for navigation/safety at sea. This work is licensed under the Web Services and API Terms of Use. For information on specific bathymetric data sources used to compile this map, links where the data may be downloaded, as well as copyrights and use constraints related to that data, refer to the Ocean Basemap.

Showcasing Success Stories in MarineCadastre.gov, created by National Oceanic and Atmospheric Administration (NOAA) Coastal Services Center (CSC); data from NOAA CSC, MarineCadastre.gov, National Geographic, Esri, DeLorme, General Bathymetric Chart of the Oceans, National Geophysical Data Center, International Hydrographic Organization-Intergovernmental Oceanographic Commission, National Geodetic Survey, University of New Hampshire, US Department of Defense, US Department of Energy National Renewable Energy Laboratory.

Tiber Reservoir Map, created by and data from US Department of the Interior, Bureau of Reclamation.

National Wildlife Refuge System, created by and data from US Fish and Wildlife Service.

US DEPARTMENT OF VETERANS AFFAIRS

VA Enrollee Population by County, created by and data from US Department of Veterans Affairs, Planning Systems Support Group.

INDEPENDENT FEDERAL AGENCIES AND TRIBAL GOVERNMENTS

How's My Waterway? created by US Environmental Protection Agency (EPA), INDUS Corp.; data from EPA.

Facility Registry System on the Exchange Network, created by the US Environmental Protection Agency (EPA) and Blue Raster; data from EPA Facility Registry Service, Air Facility System, Assessment, Cleanup and Redevelopment Exchange System, Air Quality System, Base Realignment and Closure, Biennial Reporters, Bureau of Indian Affairs Indian School, DTSC-EnviroStor, Clean Air Markets Division Business System, Clean Watersheds Needs Survey, Comprehensive Environmental Response, Compensation, and Liability Information System, Electronic Greenhouse Gas Reporting Tool, Emission Inventory System, Emissions & Generation Resource Database, Energy Information Administration, Enforcement Criminal Records Management System, Environmental Protection Computer System, Integrated Compliance Information System, National Compliance Data Base, National Pollutant Discharge Elimination System, Office of Transportation and Air Quality Fuels Registration, RACT/BACT/LAER Clearinghouse, Radiation Information Database, Renewable Fuel Standard, Resource Conservation and Recovery Act Information, State Environmental Programs, Toxics Release Inventory System, Toxic Substances Control Act, state and tribal environmental monitoring agencies.

Consolidation Opportunities, created by and data from US General Services Administration.

GSA Mapping Portal, created by and data from US General Services Administration.

GSA's Multi-Asset Planning Tool, created by and data from US General Services Administration.

GSA Major Construction Project Map, created by US General Services Administration (GSA); data from GSA, Esri, DeLorme, HERE.

Community Planning and Sustainability Planning Map, created by and data from US General Services Administration (GSA); data from GSA, Esri, DeLorme, HERE, Food and Agriculture Organization of the United Nations, US Geological Survey, US Environmental Protection Agency, National Oceanic and Atmospheric Administration.

Grex Leasing Tool, created by US General Services Administration (GSA); data from GSA, US Geological Survey.

Honolulu Transit Maps, created by US General Services Administration (GSA); data from GSA, Esri, DeLorme, NAVTEQ, US Geological Survey, US Department of Agriculture, US Environmental Protection Agency, National Geospatial-Intelligence Agency.

Voyager Card Fuel Transactions 2012–2013, created by US Postal Service Office of Inspector General (UPS OIG); data from UPS OIG, ©2012 NAVTEQ.

Office of Inspector General Relocation Map, created by US Postal Service Office of Inspector General (USPS OIG); data from USPS OIG, Esri, DeLorme, HERE, TomTom, Intermap, Increment P Corp., General Bathymetric Chart of the Oceans, US Geological Survey, Food and Agriculture Organization of the United Nations, National Park Service, Natural Resources Canada, GeoBase, Institut Géographique National, Kadaster Netherlands, Ordnance Survey, Esri Japan, METI (Japan Ministry of Economy, Trade, and Industry), Esri China (Hong Kong), swisstopo, and the GIS User Community.

Smithsonian Institution Explorer Maps, created by and data from Smithsonian Institution.

Land Use Public Comment Map, created by and data from Confederated Tribes of the Umatilla Indian Reservation.

Explanation of River Mile Discrepancies in Rice, Washington, created by University of Washington graduate students Grant Novak and Jon Walker, Spokane Tribe; data from US Geological Survey, Washington State Department of Ecology, Washington Department of Natural Resources, Spokane Tribe.